Motion Media and Communication:
The History of and Strategies for Influencing Audiences through Kinetic Visuals

S. Martin Shelton

Lamplight Press
Austin, Texas

Motion Media and Communication: The History of and Strategies for Influencing Audiences through Kinetic Visuals

Inquiries should be addressed to the Publisher:
Lamplight Press
PO Box 82516
Austin, Texas 78708

ISBN: 978-0-9979774-7-9

Printed in The United States of America at Lightning Source

Table of Contents

List of Illustrations

Dedication

I dedicate this book to my wife, Mary Allene Shelton, and my daughter, Rachel V. Shelton. Both have given me unstinting support in my career as a filmmaker and in the preparation of this book. I could not have done either successfully without their encouragement, help, and understanding.

Acknowledgments

Over the many years of my career, I've had caring mentors who have given me expert guidance and counsel and who have shared their expertise willingly and patiently. My peers and fellow professionals have challenged me to do better. Teachers whetted my appetite for this wonderful sight-and-sound profession of ours, stimulating me to constantly extend my professional development. Fellow members in professional societies, such as the Society for Technical Communication, the Information Film Producers of America, and the International Television Association, have inspired me to emulate their communication and managerial achievements.

Without the help of all these people, I would have long ago withered in this profession. It is they who make this book possible. Heartfelt thanks to one and all. In particular, I want to name the following longtime friends and mentors:

My shipmates in the Pacific Fleet Combat Camera Group in the early 1950s:
- Lieutenant (jg) Charles (Chuck) Kircher, USN (ret.). Challenged me to learn how to fashion my cinematography for the editor to assemble filmic design scenes.
- Chief Photographer's Mate Franklin (Kaz) Kuazukaitis, USN (deceased). My Leading Chief in our detachment in Korea. Encouraged me to see the story and capture it with my motion-picture camera.
- Chief Photographer Mate Patrick (Pat) Cady, USN (deceased). My team leader in Korea and at the Naval Photographic Center. He demanded that I do my very best and sparked my passion for this profession.

Other professionals:

- Melvin (Mel) Sloan, (deceased), professor in the University of Southern California Cinema Department. He taught me the basic wherewithal of our profession and instilled in me the maxim that communication is paramount in information film.
- Elizabeth (Betty) Scheibel Dunn, (deceased). My longtime home-town friend and a reporter for LIFE magazine, whose critiques and encouragement were the sparks that inspired me to see the beauty of our language.
- Erskin (Gil) Gilbert, (deceased). An information-film scriptwriter who with patience and skill guided my filmic-design skills.
- Alberta (Berti) Cox, former president of the Society for Technical Information. She demanded that my writing be sharp, sharper, and sharpest.
- Marilyn Morgan, professional editor. She was instrumental in integrating my articles, papers, and jottings into a coherent whole that became this book.
- Emilie Boguchwal and Janice Kasperson, members of the Society for Technical Communication, Sierra Panamint Chapter. Over the years, they edited my ramblings into publishable manuscripts.
- Louise Burnham, a student in my workshop, who developed the first draft of the script *The Scarf: The Perennial Fashion Statement*.
- Ken Albright, intrepid ghost town hunter and longtime associate. His insightful critiques honed this manuscript.
- Olivia Frances, commercial artist. Thanks for the *The Scarf* storyboard.
- Ron Marryott, Jr., former associate and longtime friend, for his sage counsel.
- Brian McCaleb, Consulting Professionals United, for his detailed review of this book and his singular assistance in making it ready for publication.
- Richard (Dick) McCurdy of McCurdy Editorial for his review of and comments on the chapter "The Sound Track: An Integral Part of Information Motion-Media."

- Mark Pahuta, former associate and longtime friend, for his comprehensive review of and cogent comments on the first draft of this book.
- Margaret Dalrymple, editor who honed the manuscript into a smooth-reading text.

Thanks are also owed to the following for their gracious permission to use materials reproduced in this book:

1. The National Film Board of Canada
2. The *Bildarchiv* of the *Deusch Presse Agentur* (Archives of the German Press Agency)
3. The Margaret Herrick Library of the Academy of Motion Picture Arts and Sciences
4. Larry Edmunds, Inc.
5. United States Air Force
6. UCLA Film Archives
7. Mrs. William Wyler
8. National Audiovisual Center
9. Museum of Modern Art, Film Archives
10. Sunbreak Productions
11. WTTW, Boston
12. Hollywood Books
13. United States Navy

Preface

I was active in the information motion-media (kinetic sight-and-sound media) profession for almost all my adult life, and I loved it. I'm truly fortunate to have found my life's calling in a vocation that has posed interesting challenges every day—I always learned new things, met interesting folks, and worked with consummate professionals (on the whole). I have traveled the world several times over and had a great time doing it.

Propitiously, the Hollywood movie bug never bit me. I was always focused on the information motion-media rather than the narrative film, the "movies." My fiscal rewards were adequate, but the real reward was the love and satisfaction I got from working in this wonderful profession. Perhaps more important, my reward was the pride and fulfillment I experienced seeing my shows on the screen. Watching audiences react positively. Garnering praise from clients. Earning awards in festivals. And even dealing with my share of "learning experiences." On the whole, it doesn't get any better than that.

During my long career, I was fortunate to have attended fine schools and to have had demanding teachers, great mentors, exceptional professional associates, empathetic managers, tolerant clients (mostly), an understanding family, and the God-given wherewithal to learn and to work with industry and verve.

Eventually, time caught me. Several years ago, I retired from active script design, production, and management. On reflection, I'd do it all over again—perhaps a few things differently but essentially the same. Nowadays, I'm writing historical adventure novels, short stories, and a technical monograph, and having a fine time in my golden years.

Actually, this book on filmic communication has been a work-in-progress for some thirty-five-odd years. It's a synthesis of material I've written as papers for professional journals, magazine articles, editorials, newsletter columns, jottings, and scribbles on notecards. This book isn't organized like a textbook or a how-to book. It's a series of essays on motion-media that I've edited and arranged into a coherent pattern in five modules. Each module deals with one major topic. The book can be used as a handy reference for quick perusal to get the kernel of an idea on a particular topic, or it can be used as a "sit-down" study in which all the ideas are set in perspective for in-depth thought.

Frequently, I cite old-time classic films rather than contemporary ones. The pioneers and giants of our profession produced these films with communication skill, adroit filmic design, and artistic finesse—and there were no digital effects (Amen!). I'll concede that most of these films are old, timeworn, and out of style; nonetheless, they set the standards for our profession.

The primary reason I'm writing this book is to speak my mind in a forthright manner on the status of our profession and to encourage newcomers and old-timers alike to consider adapting the ideas I'm proffering to enhance their professional growth and our profession. Another reason is to fulfill my obligation to leave a legacy to my profession.

I step on many toes in this book and debunk a host of "sacred doctrines." There is some "how-to" in it, but more than anything else, this book is an editorial. By nature, I'm a contrarian, so please bear with me. I'm trusting that your patience and understanding will be rewarded.

Prologue

I'm sorely disappointed in the state of our profession—the information motion-media profession. We're not doing very well. Far too many of our films, videos, and multimedia shows are amateurish in concept, technically inept and, most importantly, fail in their primary objective of communicating ideas to our audiences. Unfortunately, many of today's films, videos, and multimedia shows sparkle with "creative" digital effects and whiz-bang-and-pop multichannel surround-sound, but they don't communicate much except, perhaps, "Gee! Isn't technology wonderful?"

Nowadays, we no longer have to think very much about it. Rapidly advancing technology has made it remarkably easy for anyone with an iPhone and a simple computer to produce a motion-media show. Throughout this narrative, I'm using "show" to refer to the generic, all-encompassing information motion-media tool.

We've lost track of what we're up to. We're not supposed to be producing films, videos, television commercials, and multimedia shows; we're supposed to produce motion-media shows that communicate ideas to our target audiences—shows that implant our messages firmly in audiences' minds so that when "the lights come on" they will think, act, or speak in the manner that fulfills our show's goals. Our show's goals need be *fitting, realistic,* and *worthwhile.*

It's the communication achieved that matters. All else is irrelevant. We use kinetic-visual media as the carriers of our encoded messages. It's the transmitted message and our audience's reception that are germane. That's what this book is about: how to optimize communication using motion-media—integrated sight-and-sound kinetic media.

This book is not about technology or tools. There's not one comment about f-stops, key lights, chrominance, bit rate, gamma curves, white balance, decibels, script writing, auteur concepts, or the Strasberg acting method. If you're looking for the latest technical how-to, this tome is not it. There are plenty of excellent books on these specialized subjects.

Rather, this book takes an in-depth look at just what our profession is about: **communication**. The messages in this book reflect my perspective—part theory, part psychology, and part philosophy: its analysis, history, application, and personal opinion. I approach the topic from a reality-based understanding of the media.

I explore the entire genre of *information* motion-media whose purpose is to inform rather than to entertain.

Occasionally, I'll discuss experimental, enrichment, or personal-statement shows. Specifically, I've excluded narrative motion-media—the "movies."

Fully realizing that I'm challenging some of our media professionals' sacred tenets and widespread practices, I stand pat. My opinions are based on a long-term and successful career in all aspects of motion-media production, and on viewing dozens of motion-media shows a year. For thirty years, I was a preliminary and blue-ribbon judge in several national sight-and-sound media competitions. For twelve consecutive years, I managed a major international motion-media competition for one of the nation's largest professional communication societies. I made it a point to view shows of various genres from other competitions and from government, industry, and a deluge of television commercials. Unfortunately, I've seen distressingly few shows that match the communication beauty of *To Be Alive* by Francis Thompson, *Why Man Creates* by Saul Bass, *Man with a Thousand Hands* by Cap Palmer, *Universe* by Lester Novoros, and dozens of others that engender effective communication.

Why is this so? Why are we inundated with "good enough" mediocrity and wholesale incompetence in the vast majority of the shows? Where are today's Thompsons, Basses, Palmers, and Novoroses? One possible answer is that many of today's media designers and producers don't know, don't care, or don't understand that they're in the communications profession rather than the entertainment industry. They are seduced by the false notion that they must entertain and be creative with digital wizardry to get the audience's attention. [See Figure 5, Chapter 4 (the Bo-10), page 30, for an all too real example.]

I'm not suggesting that our shows must be dull, lack innovation, or not excel technically. I am suggesting that we need to analyze critically the psychological, technical, and artistic elements planned for our shows and to evaluate their contributions to its communication power. All factors being equal, we optimize the power of our communication medium when our messages are contained in kinetic visuals and edited into montage scenes—the relevant filmic design.

Part I: Introduction

Chapter 1

The Message and Motion-Media

We're in the cyberspace world of interactive, digital communications. We send and receive many of our motion-media messages through the ubiquitous Internet, the Worldwide Web. The technological explosion in electronics and computers has transformed our profession from a mostly handcrafted profession into a technological specialty. It's not out of the ordinary to do photography in one medium, edit in another, record and dub sound in another, and release the show in another or in several media.

Nonetheless, all-encompassing motion-media has many forms and uses, ranging from theatrical extravaganza to lyric poem, from hard-hitting documentary to the fluff of a television situation comedy, from training aviators to experiencing the abstractions of Norman McLaren.

In today's high-energy motion-media environment, we are inundated with a myriad of multifaceted messages, all competing for our time, energy, and attention. The barrage never ceases. Alvin Toffler puts it this way: "The waves of coded information turn into violent breakers and come at a faster and faster clip, pounding at us, seeking entry, as it were, into our nervous systems."[1]

I suspect that we've become too passive in this media-rich environment. It's difficult for many of us to sort and evaluate information, to visualize or conceptualize ideas and philosophies, to reason. We are beclouded in communication fog. For instance, Everett L. Jones of the University of California, Los Angeles, found that steadily increasing numbers of students beginning college cannot write a simple declarative sentence. They must be enrolled in remedial, noncredit writing classes, with many taking the class several times to earn a passing grade.[2]

This complex problem is summarized succinctly by the late Mike Wallace of the Columbia Broadcasting System, who wondered "… if we all aren't vastly over-communicated, if words [and television pictures] haven't become a kind of Muzak® for us, a background hum that fills the silent gaps of time in which otherwise we might just sit and think."[3]

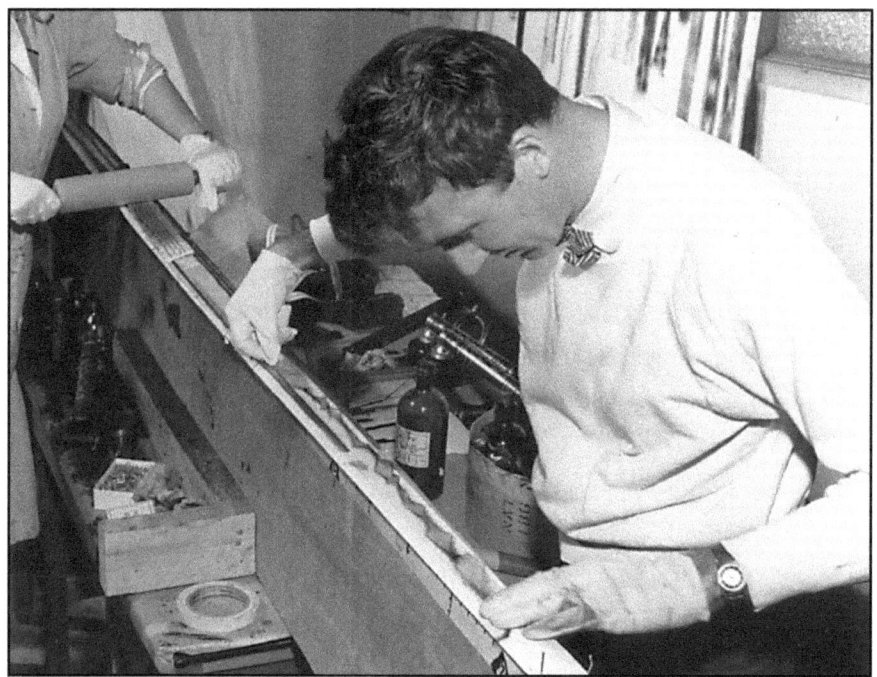

Figure 1. Norman McLaren drawing directly on 35mm film for a scene in his film *Begone Dull Care*. Photograph courtesy of National Film Board of Canada.

Communication Business

Our task is to ensure that the messages we transmit via motion-media are not Muzak and do not numb the senses. Accordingly, it's the communication business we're in, not the film, video, or multimedia business. Understand this precept to the depth of your being. Actually, when we strip our profession to its essence, we're in the psychology business. Our task is to manipulate the minds of our audiences so that when the lights come on they will do, say, act, or think according to the goals we have set forth in our shows.

Shelton's Pronunciamento

We are in the communication profession, not the film/video industry.

Here's a potpourri of the general goals of our shows: to arouse, document, educate, indoctrinate, influence, inform, inspire, motivate, orient, persuade, propagandize, report, recruit, seduce, teach, train, or sell.

Let's explore motion-media's general characteristics as a *mass communication* tool. A motion-media show:

- Communicates messages to large, heterogeneous, and anonymous audiences[4]
- Communicates the same message to mass audiences simultaneously, sometimes in public, other times in private
- Embodies messages that are usually impersonal and transitory
- Is multi-sensory in that the audience's sight and hearing are stimulated in concert
- (Such a combination of sensory stimuli forms a complex synergism that significantly enhances communication.)
- Is a formal, authoritative channel of communication[5]
- Is instrumental in behavior and attitude modification
- Confers status upon issues, persons, organizations, or social movements[6]

Kinetic-Visual Communication

We are a visual species. Television, motion pictures, videos, graphic novels, picture magazines, multimedia, and the Internet are just some examples of the visual media to which we are exposed. It's common to find information conveyed not through words but through icons, for example, on computer menus, restaurant bills-of-fare, and traffic signs.

The point is, we don't think much in words themselves. Words are too linear and too slow. Admittedly, words can create visual images, but it's only by convention that they represent their referents. Vision is our primary medium of thought, and graphic images are the most powerful way of enhancing our perceptual thinking.[7] Marshall McLuhan's adage "The medium is the message" prevails.[8] I disagree. I'm convinced that "The message is the message!"

We use our sight-and-sound, kinetic media to communicate relevant messages that stick in the minds of the target audience and cause them to achieve the goals that are set for our shows. In other words, in order to communicate effectively, we must encode our messages into a filmic design that engenders maximum audience empathy. Empathy is that close identification an individual has with a person, place, event, institution, or some combination of these items. It's the psychological and emotional involvement the individual has with a stimulating situation. Skillfully designed motion-media shows are especially adept at engendering intense empathy. A number of researchers have verified the thesis that empathy is the primary constituent of communication in information motion-media; in general, the more intense the involvement, the more effective is the motion-media show.

Shelton's Pronunciamento
Empathy engenders communication.

There is a caution, however. Some research indicates that, should the empathy be too intense for an individual, communication effectiveness may decrease markedly. Because of an intense emotional experience, a viewer can make a near-total projection into the "reality" of the scene, especially if the scene mimics a phase of their lives that is, or is perceived to be, <u>threatening</u>.

For example, (many eons ago) for my Master of Arts degree in Cinema at the University of Southern California, I conducted a research study on the effectiveness of humor in a training film. I produced two nearly identical "how to do it" shows. One demonstrated five key steps in the process straightforwardly. In the other film, I couched these five steps in a humorous *milieu*.

- My premise was that the target audience had limited skill in changing the wheel of an automobile. Written and physical pre-testing proved my premise.

- The target audience were female, high-school junior and senior students enrolled in driver education classes. Before screening and through written assessments, I had identified my audience through several criteria: intellect, mechanical skill, and attitude.
- The films' goal was to teach these students how to change the wheel of an automobile when the tire is flat (a perceptual-motor skill).

After screening, I gave a written test to determine the students' immediate learning. One week later, my associates and I administered performance tests: students changing a wheel.

The students that viewed the straightforward film did exceptionally well in the written and in the performance test.

My committee and I got a unexpected surprise. Many of the students who viewed the film with the humor *milieu* failed the written and practical tests. I reviewed their comments on the critique sheet and found that the empathy engendered in this audience was *too* intense. The students had projected themselves wholeheartedly into a scenario perceived as threatening. The situation was too serious to be treated humorously. The girls expressed fear of being alone and in such a predicament that they had to perform this task for their safety. Point taken.

****

Viewing is the key. The human visual system is an incredibly powerful receiver. For most of us, audible speech plays a strictly subordinate role in the reception of information.[9] And speech should play a similarly subordinate role in our films, videos, and multimedia shows. Some of the best shows I've seen over the years are those that engender maximum communication—they have no, or only minimal, narration or dialogue. Often I have seen such shows win top awards in competitions. For example,

- Albert Lamorisse's *Le Ballon Rouge (The Red Balloon),* Palme d'Or, Cannes Film Festival, 1957
- Frederic Back's *Crac!,* Academy Award, 1982
- Bob Rogers's *Ballet Robotique,* Best of Show, Information Film Producers of America, 1982

Education, Educators, and Advanced Technology

Let's take a sidetrack for a few lines of text. I want to discuss a pet peeve.

A while ago, a friend of mine, a professor of cinema/television at a prestigious university, asked me the following questions: "How can educators best teach advanced technology in information film and television production? How can educators develop their own expertise in this technology?"

From my perspective, the professor asked the wrong questions. His emphasis was first on learning advanced technology, and then on teaching it. What equipment is the industry using now? Which buttons do we push and what levers do we pull to get the "XYZ" effect? How do we integrate such technology to produce motion-media shows, and the host of other such media sure to follow?

These questions and the idea posed in the initial premise are valid. However, they are secondary. Such a perspective puts the cart before the horse, as it were. The professor's emphasis was on the technology rather than on the more fundamental aspects of communication. His questions ought to have been: How do we teach our students to encode information into coherent messages effectively and efficiently with the new communication technologies? How do we transmit such messages so that ideas are firmly implanted in the target audiences' minds and affect their behavior in a manner that meets our communication objective? *All else is irrelevant.*

I understand what's happening. We are seduced easily by advancing technology that apparently makes our jobs easier, faster, and more creative. We're bedazzled by the razzmatazz. Technology, per se, overwhelms us, and its use becomes our primary goal. Communication becomes secondary. I'm convinced that today's educators, students, and media professionals don't understand that motion-media is a communication tool and not an end unto itself.

If we focus our endeavors on teaching university students the technical aspects of our profession, we've become a technical training school instead of an academic institution. For now, it's of minor significance that students know which buttons to push and levers to pull. Besides, technology is changing so rapidly that it's impossible to train our students in the "latest."

It's through on-the-job training that former students become proficient in using ever- changing technology.

The university is where ideas, concepts, and rational thinking must be paramount. It's far more important that motion-media students have an understanding of communication theory, psychology, logic, research, communication skills, and system analysis. At least, that's my gospel and what the rest of this book is about.

Shelton's Pronunciamento
Technology should enhance communication value, not overwhelm it.

Chapter 2

Motion-Media in Information Society

Communications in the Twenty-First Century

Technology has developed so rapidly that we can develop complicated motion-media shows with our hand-held devices and personal computers—shows that just a few years ago required a multi-million-dollar studio, complicated processing laboratories, and technical and animation facilities to produce. It's almost too easy. Has technology seduced us into forgoing our basic communication skills? Are our shows technically excellent but communication-poor? Has technology exceeded our ability to use it to communicate effectively?

Communication in our information society focuses on mass audiences and on interactive, personalized requirements. It's diversified and tailored. There is a plethora of niche television channels available today that are targeted to a vast variety of special interests. We'll not have a problem filling the multichannel interactive television (or whatever) system with programming.

It's beyond my imagination to envision the advanced communication production and distribution technologies that will evolve as this twenty-first century develops. Here are a couple of such technologies that I know are currently in research and development:

Kinetic holograms. Nicholas Negroponte, former director of Massachusetts Institute of Technology's Media Lab, envisions the reproduction of a real-time, three-dimensional, high-fidelity, surround-sound, kinetic hologram of exceptionally high quality. I heard him say, "We could 'experience' the Super Bowl being played from end to end in our media room—the ultimate telepresence scenario."[1]

Scalable TV. This technology will allow viewers to determine the size of the picture they screen. No matter the size of the picture, the image quality will be the same. It's a matter of the number of pixels (information) per unit area on the monitor—the larger the image, the more pixels on the monitor.

Media Responsibilities

Motion-media is society's primary source of information—integral to informing citizens about government, business, society, international affairs, etc. Its primary functions are to:

- Report straight news and features
- Make editorial comment
- Disseminate propaganda
- Comment on societal activities
- Document society's heritage

Considering these powerful functions of motion-media, those in control of the media have an ethical responsibility to the public and to their profession to be honest brokers.[3] Unfortunately, the media has its cadre of communicators and opinion leaders who bias mass-media communications with their own orthodoxy.

Should a communicator or opinion leader control the mass media to the degree that the populace has no access to alternative information, the communicator has effective control of the social and political environment through manipulation of the behavior and thoughts of the proletariat. Such is the case in despotic regimes.

Significantly, my longtime friend, now deceased, Father Louis Reile, Society of Mary, professor of film at St. Mary's University in San Antonio, said, "The prophet has a new tool, the cinema, that graphic art of reality caught on celluloid and shot through a magic lantern to simulate life."[4] Is the prophet ethical or immoral? To what end is this "magic lantern" used?

Fortunately, we can tune in dozens of worldwide television channels beamed down to us from an array of satellites. We can even broadcast our own comments over private networks.

The Communication Challenge

I'm convinced that kinetic graphics with associated sound are the most powerful way of enhancing our perceptual thinking. Rudolf Arnheim says, "Perceptual and pictorial shapes are not only translations of thought products, but they are the very flesh and blood of thinking itself."[5]

Let's continue to produce and expand significantly mind-challenging motion-media shows that communicate science, geography, language, history, and ideas of all types. It's been done long before and most effectively by the masters of our profession. For example, Leni Riefenstahl used film with telling effect in promoting the German National Socialist (Nazi) Party in the 1930s, in *Triumph des Willen* (*Triumph of the Will*) and *Olympiad*. Conversely, Major Frank Capra, US Army, used film with striking power in the *Why We Fight* series to rally American morale and efforts during World War II—yes, the same Frank Capra who produced *It Happened One Night*, *Mr. Smith Goes to Washington*, and a host of other people-oriented films.

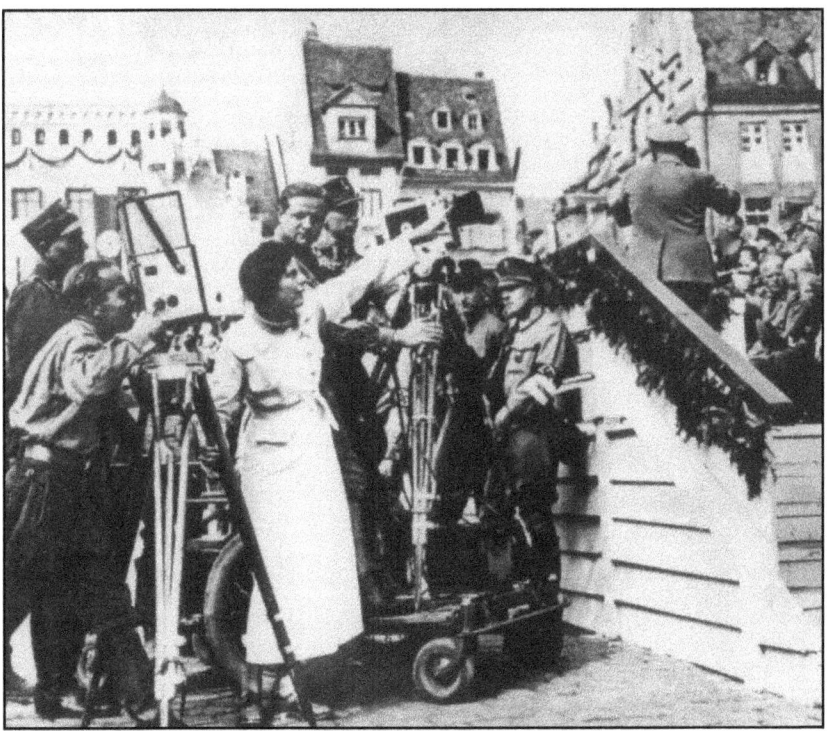

Figure 2. Leni Riefenstahl directing a scene for her film *Triumph des Willens*.
Photograph courtesy of German Press Agency.

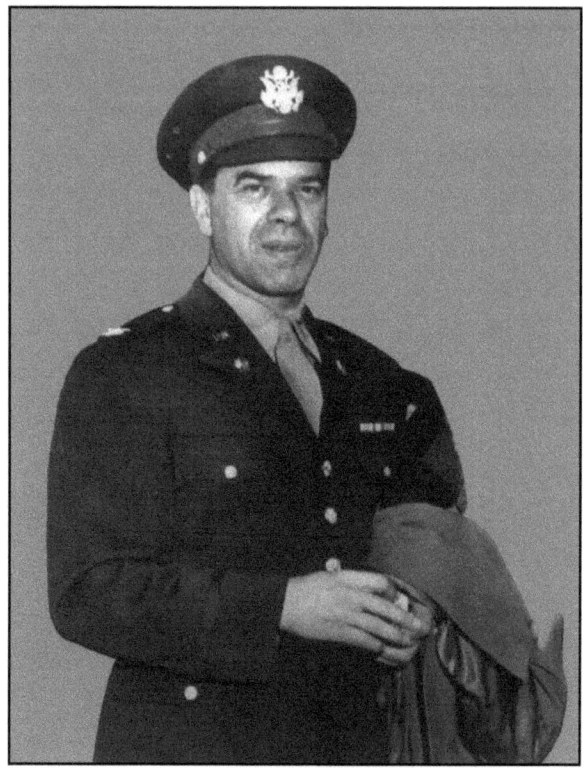

Figure 3. Major Frank Capra in 1943, producer of the highly successful nine-film series *Why We Fight*. Photograph courtesy of the Academy of Motion Picture Arts and Sciences.

Medium Versus Message

Frequently, a discussion on message encoding into media stirs controversy about whether the medium or the message is the more important factor in communication. This question is timeless, much debated, and unresolved to any degree of satisfaction. Because the thrust of this question is germane to the coding techniques required to transform the essential elements of information into motion-media, a few comments are in order to establish a perspective.

Marshall McLuhan, discussing how electric light was needed for activities such as night baseball and brain surgery, makes the point that "the medium is the message." He reasoned that brain surgery and baseball could not be accomplished without light and, in some ways, are the "content" of the light because it is the medium that controls all human activity.[7]

Conversely, there are a number of researchers and communicators who espouse the concept that the message is the more important factor in communication. For instance, C. F. Hoban and E. B. Van Ormer, after an exhaustive and scholarly review of the research literature, proposed that there is nothing in the medium of film per se that guarantees communication. Communication is a direct function of how well the messages relate to the audience.[8]

Consider, for example, the following scenario. A flying saucer lands on the White House lawn. Little green men hop out, storm the White House, capture the president, and whisk the chief executive away in their saucer. If this event were captured on a recorded medium, no matter how technically incompetent the images—out of focus, overexposed, or scratched—this scene would be a record of incalculable value.

A real-life example is the 22 November 1963 Abraham Zapruder film that recorded the assassination of President John Kennedy. The slightly less than perfect, silent, 8mm color motion picture sequence that captured this historical event is paramount. All filmic design and technical excellence fade to insignificance.

Shelton's Pronunciamento
The importance of the message overwhelms all filmic considerations.

We have seen "terrific" information motion-media shows that have dazzled us with their brilliance and showmanship, and we were much impressed. Yet, on reflection, we realized that the shows did not say much to us. Perhaps the shows entertained, but they did not communicate the intended messages. In these cases, the medium is the message, but entertainment is not the reason the audience viewed the shows. Also, the unsophisticated and unwary may be enchanted by a medium—the "silver-screen magic" of a film, for example. If a show is produced, some viewers will conclude that "it must be important."

The simple fact is that too many of us have focused on the selling of media rather than on relevant messages. Some of us are more concerned with the gadgetry, hardware, software, and glamour of it all than with communication concepts. This is not to discount these items; without them, we could not produce our shows.

Wrap

The processes and techniques available to the filmic designer in structuring and producing our information motion-media show are many, varied, and powerful. They should be used judiciously toward accomplishment of communication goals.

In the future, it matters not what communication tool we use—media, medium, multimedia, hypermedia, supermedia, or something else. What matters is that we use our communication skills and technology to design and produce shows that communicate our messages to the target audiences readily, efficiently, and economically.

Chapter 3

Some Thoughts on Information, Communication, and Meaning

We're in the communication business rather than the film, video, or multi-media business—if you don't agree, put this book down and watch a rerun of *Gilligan's Island*. Now, let's define terms, explore concepts, and develop an understanding of what this profession of ours is truly about. We'll explore briefly a few communication fundamentals and look at Claude Shannon's information theory. An appreciation of such fundamentals (even if not in depth) is essential to the comprehensive understanding of my ramblings in this book.

Message, in the broad sense, is whatever stimulates our senses. David K. Berlo defines messages as "(an) ordered selection of symbols intended to communicate information."[1]

What's relevant is how our audience perceives our messages. Consider whether your messages are
- Real or fiction
- Objective or biased
- Believable or incredible
- Informative or abstruse
- Logical or irrational

Also, the style in which we encode our messages has significant influence on our audiences. Message style can be

- Artistic or base
- Stimulating or boring
- Concise or discursive
- Relevant or gratuitous

These perceptions, coupled with the show's filmic design, define our motion-media's effectiveness.

Meaning

Meaning is a psychological function that involves cultural, social, and environmental factors. Though meaning is contained in the messages, it's the sender and receiver that give meaning to messages through their common understanding.[2] Areas of concern regarding meaning lie in the ambiguity of the messages themselves. At a minimum, there are three interpretations of a message:

- Intended Message: what the sender meant to communicate
- Transmitted Message: what was actually sent and received
- Understood Message: what the audience perceived the sender sent

There is a reasonable probability that the audience will not understand precisely any message we send via motion-media. All too frequently, the discrepancy between what we sent and what our audiences understood is frightening. No matter how well planned, well structured, and well produced our motion-media shows are, we'll never be successful totally in our communication—because of psychological and/or technical factors.

Information Theory

Since we're in the communication profession, let's review two of the relevant theories that underlie our vocation.

Information theory evolved from Ludwig Boltzmann's classic second law of thermodynamics. Published in 1884, this law states that in any closed system its disorder increases. The system's ability to do work, to communicate, or do anything is reduced—its entropy increases.

Based on Boltzmann's second law, Claude E. Shannon and Warren Weaver, of AT&T Bell Laboratories, devised a general mathematical communication model to explain the special problems of information processing, transmission, and reception attendant with electric and electronic systems such as telegraph and telephone, and with the burgeoning electronic data-processing machines.

Shannon's theory was published in a 1948 Bell Laboratories paper entitled "A Mathematical Theory of Communication." In his theory, the term *information* has a mathematical definition specifying the probability that any given bit of information will be transmitted and received undistorted. Shannon said, "The fundamental problem of communication is that of reproducing at one point either exactly or approximately a message selected at another point."[3]

Shannon defined communication as the encoding of information into electronic signals, transmitting and receiving the signals, and decoding the signals back into information that is error-free or nearly so. In broad terms, Shannon's information theory applies to any general communication system—including motion-media.

The General Communication System

I've designed Figure 4 below to illustrate the information flow in Shannon's theory. At "Information Source," I've depicted the signal's flow in dark black to indicate that it's pristine. As the signal progresses through the system, noise continually contaminates it. When the audience receives the message, its fidelity is seriously eroded.

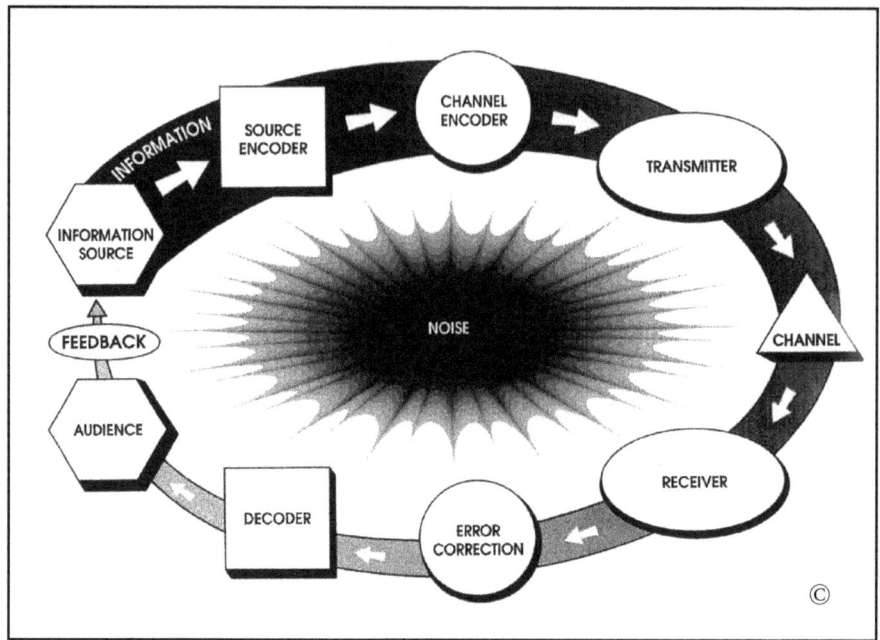

Figure 4. The Generic Communication System.

To demonstrate, let's follow a video signal as it flows through the communication system.

- Information Source: the show's sponsor, research material, or interviews with the intended audience
- Source Encoder: script designer translates the relevant information into a script and storyboard that detail the filmic design
- Channel Encoder: video camera converts visual and aural messages into electronic signals
- Transmitter: device that sends the electronic signals into the channel
- Channel: carrier of the electronic signals (wire, fiber optics, air)
- Receiver: device that captures the electronic signals
- Error Correction: device or algorithm that modulates the noise that has suffused into the system
- Decoder: component that translates the error-corrected electronic signals into usable visual and aural information—the message

- Audience: folks whose minds we're trying to manipulate (or, could be a machine)
- Feedback: audience acknowledging that they have received and understood the message
- Noise: mechanical or psychological bugaboo that distorts the message

In any communication system with a substantial degree of complexity, the probability of error infecting the system is high, whether through system failure or human frailty, the barriers that make effective communication tenuous even under ideal conditions. Three functions in the communication process are particularly noteworthy.

Noise is anything that interferes with the reception and understanding of the signals. Because noise distorts the signal, it also distorts meaning. Noise is either mechanical or semantic: static on the radio or misinterpretation of the signals because of referent differences or psychological distractions. Controlling semantic noise is, in a sense, what this book is about.

Channel is the carrier that transmits the encoded signals.

The channel for:

- face-to-face speech is vibrating energy traveling in air (sound waves)
- printed words is inked signs on paper
- video is electronic signals traveling through the air or over cable
- motion pictures (traditionally) are images and sound recorded on a strip of film
- multimedia is electronic signals

Feedback is the receiver's signal to the sender that authenticates receipt of the message, that the message is understood, and confirmation that the receiver will execute the action item in the message. Such a signal closes the communication process.

A positive acknowledgment to the sender is accomplished by such utterances as "Aye, aye, Sir," "Will do," "Okay," "I understand," "Ten-four." Or conversely, "Say again," "I don't understand," "Please rephrase."

Ideally, feedback ought to be in real time, but except for one-on-one communication (face-to-face, telephone, etc.), it seldom is. Feedback in film and video is usually very low. However, in multimedia, real-time feedback is a salient advantage. I would suggest that feedback is perhaps the <u>most important characteristic</u> of the entire communication process.

Shelton's Pronunciamento
Feedback engenders successful communication.

Information Theory Redux

In essence, Shannon's theory deals with three primary and interrelated concepts: *uncertainty, entropy,* and *redundancy.*

Uncertainty is the amount of information the receiver <u>does not have</u> about the incoming information. Shannon notes that all information we receive is a combination of what we know and don't know. The incoming information is a blend of the probability of the expected and the unexpected, the certain and the uncertain. The more information the receiver has about the incoming message, the less communication is achieved. Thus, message contents that are known to the receiver before the message is sent convey no information. Conversely, some degree of information is conveyed if an unexpected message is received.

In other words, if a person knows that the first thing the boss says each day is "Good morning," and, sure enough, that's exactly what the boss says on Monday morning, no verbal communication is achieved because the person (receiver) knows in advance what the boss's incoming message is going to be. The boss has transmitted no new information. (We're assuming that no, or minimal, noise interferes with the communication channel.)

If, on the other hand, the boss says, "You're fired!" then the information received is optimum. It's the unexpected (what we do not know), the newness or surprise, that influences the amount of useful information we receive. Thus, *communication is the resolution of uncertainty.* Such resolution is the inverse measure of entropy. That is, maximum uncertainty corresponds to minimum entropy.

Entropy is a universal law that applies to communication, thermodynamics, and all other systems in this universe. Entropy is an index of the uncertainty of a system. The higher the disorder, the greater the entropy.[4] Put another way, the higher the entropy of a message the less communication it can accomplish. The information is there, but it's inaccessible.

Ludwig Boltzmann posited in 1872 that the tendency of any closed system when left alone is to become less orderly: entropy increases to irreversible disorder. Or, things go from good to bad, maximum potential to minimum potential, order to disorder. Without external energy, all systems migrate to disorder. On the other hand, order usually is temporary and requires external energy to maintain itself once the migration to disorder has begun.

Let's suppose, for example, that we have a perfectly insulated container with a divider splitting its volume in half. Our container is totally free of any outside influence (a closed system). One side contains hot water. The other side has cold water. This thermodynamic system is orderly, and its entropy is low. The system's energy is available to do something—work of some kind, for example. Unfortunately, our divider is water-soluble. Soon the divider dissolves, and the hot and cold water mix. Eventually, all the water in the container is thoroughly mixed and is at a uniform temperature. The hot- and cold-water "system" is now completely disorderly, and entropy is at the maximum. The system can do no work or accomplish anything. Consider, then, that all worthwhile accomplishment results in an irreversible increase in entropy.[5]

You may well ask, "What's this got to do with communications?" Excellent question. Before a communication system transmits a message by speaking, writing, or projecting, it has the ability to expend energy to send a message. Consider our example of the boss's Monday morning utterance. In the first instance, entropy is high. No communication is accomplished because the person knew in advance exactly what the incoming message was. In the second instance, entropy is low. Maximum communication is achieved because the person had no clue that "You're fired!" was to be the incoming message.

The higher the entropy, the lower the communication potential; and the lower the entropy, the higher the communication potential. This maxim is valid because <u>all</u> possible messages are equally likely to be sent. Of course, after the person receives the message and the message is understood, entropy is high again. Nothing more regarding the communication can be accomplished. In effect, noise prevails.[6]

Redundancy is the transmission of more messages than are needed to ensure maximum communication; it's the predictable or conventional in a message. In almost all transmissions, some redundancy is necessary for feasible communication. In the broad sense, redundancy is the set of rules by which we encode the information we are transmitting. We use redundancy to achieve a high probability of precision communication.

For instance, our English language has "silent" letters in many words. Because we are knowledgeable about English abbreviations, we would most likely understand the word *letter* if it were written *ltr.* Communication is achieved because we know our language and know that certain sequences of letters are more likely to be correct than others. To enhance redundancy in radio communication, we use the NATO phonetic alphabet. For example, we use the word *Alpha* for the letter *A*, *Quebec* for the letter *Q*, and *Zulu* for the letter *Z*. To further increase communication, all words in this alphabet have two symbols and the first symbol is stressed and pronounced hard. For example, *Alpha* is pronounced "AL fah," *Quebec* is "KEH beck," and *Zulu* is "ZOO loo."

Another reason we use redundancy is to open channels of communication. For example, we might say on meeting our friend for the first time that day, "Hi, Joe! I met Mary last night." "Hi, Joe!" is phatic communication. It's the opening gambit. It contains minimal communication, but it's important because it alerts our receiver to pay attention: the primary or content message is coming. I call this phatic function the "heads-up." The content message is "I met Mary last night." We must assume that the meeting last night has meaning to both parties and that they know Mary. Sender and receiver have a common referent, and communication is accomplished—assuming noise in the environment is insignificant.

Our task as communicators is to keep entropy low, to reduce our audience's uncertainty by sending, in the appropriate channel, properly encoded messages that are meaningful, understandable, and, on the whole, new. We'll discuss these concepts in some detail in later chapters.

Shelton's Pronunciamento

Uncertainty, Entropy, and Redundancy are the keys to communication.

Shelton's Theory of Communication

We'll conclude this chapter with a discussion of my theory of communication. I've been in this profession for a long, long time, and I've come to understand that there's another set of interrelated criteria that influence all <u>human</u> communication systems. These criteria define the ease or complexity of any communication task. Listed in descending order of relevance, these criteria are:

- Importance. How important is the information to the audience? The more important the information, the easier the communication task.
- Is it to the audience's advantage to know the information? For instance, if the audience knows that to get promoted they must demonstrate skill in operating the Acme Thingamajig, they'll probably work our multimedia program that explains the operations of the Acme Thingamajig to a fare-thee-well.
- Urgency. When must the audience use the information? When must they take the required action? The more urgent the information, the easier our communication task. For instance, a broadcast message that tells us that a Category 5 hurricane is imminent requires prompt action. Conversely, an announcement that our airplane is two hours late requires no immediate action—unless such lateness has immediate significance.
- Currency. When does the value of the information expire, become nearly worthless, or become historical? The more current the message, the easier the communication task.
  - When is the information of little or no value to the target audience; when does it become obsolete?

- Is the information timeless?
- Motivation. Why is the audience interacting with the media: viewing, listening, and reading?
  - Is the audience interaction voluntary or compelled?
  - Are the folks bored or interested in the subject matter?
- Anticipation. Audience communication is enhanced the more the audience knows about the upcoming media event.
  - Has the audience been alerted with a "heads up," with oral or visual announcements?
  - Has the audience done preparatory work?
  - How much background information does the target audience have on the essential elements of information?
  - How much lead time will they have to prepare emotionally and physically for the event?
- Predisposition. The audience's mindsets are critical elements in the reception of our messages.
  - What is the target audience's disposition toward the sponsor of the show? The communicator? Information? Medium?
  - What are the audience's prejudices? Are they sympathetic, neutral, or hostile? To what extent are these prejudices manifested?
- Distribution. What medium or combination of media do we use to transmit the information in order to optimize audience communication: speech, newspaper, motion-media, radio, online?
  - Some of the factors affecting the distribution medium's influence on the audience are their familiarity with the medium, preferences, access to the medium, and trust.
- Environment. Is the ambience of the event's environment conducive to the communication task?
  - What are the effects of the physical surroundings: ambient noise and light, climate, décor, etc.?
- Feedback. Does the audience have the opportunity/option to tell us how we're doing?

- Does the target audience participate in the learning process through feedback?
- How important is audience feedback in accomplishing the show's communication goal? Is it required? Desired? To what extent? Not relevant?

With these criteria optimized, and if all other communicator factors are favorable, we've an excellent opportunity to engender empathy in our target audiences. On the whole, *the higher the empathy, the greater the communication potential.*

I fully realize that the concepts I've discussed in this chapter are complicated and probably unfamiliar. However, they are the essence of our communication profession and apply equally well to all media—personal and machine. It's imperative, therefore, that we understand these concepts and know how to apply them in our script-designing and motion-media production efforts.

Chapter 4

Creativity May Not Equal Communication

The Alchemical Elixir

Our goal must be to design and produce successful motion-media shows, but few of us ever do it, certainly not on a continuing basis. Our shows are often less successful and acclaimed than we intended because we've deduced erroneously that a great show is a natural result of our application of creativity. To many of us in the communication profession, "creativity" has become a magical potion that ensures a "sockeroo" show. It's a panacea that solves all communication problems. All too often, I've heard folks in our profession discuss creativity with a reverence that approaches idol worship—it's the alchemical elixir that cures all communication problems. We seem to be enraptured with the ever-expanding palette of advanced technology that enables us to produce shows with a dazzling degree of sophistication. Such sophistication makes a "great show," I've heard.

Such sophistry seduces us. I'm convinced that creativity becomes the *raison d'être* of many of our shows—communication fades to a secondary purpose. Our clients and audiences are disappointed because their communication goals are not realized. The reason is that we do not understand what creativity is, and we misconstrue its worth in motion-media communication.

One example of this pervasive attitude that infects our profession was in our federal government's audiovisual (motion-media) standards, as published a few years ago.[1] To qualify to bid on government motion-media (audiovisual) contracts, prospective producers had to submit a sample of their shows for evaluation by a panel of "experts." On several occasions,

I served as one of the experts on such a review panel. The shows we reviewed, overall, were inept—void, or nearly so, of communication values and technical prowess. Significantly, many of the "experts" on these panels did not know the difference between a sprocket hole and an f-stop. One was a secretary with no motion-media training or experience. Trying times, indeed!

Out of a total possible score of 100 (70 is the qualifying score), a show could earn up to 20 points for "creativity." And creativity was divided into two subcategories:

1. "Was the show fresh and innovative? (0 to 15 points)"
2. "Was the manner of presentation appropriate? (0 to 5 points)"

Our federal bureaucrats did not define "creativity." Nor did they state why creativity ought to be included in the scoring. Rather, the bureaucrats had concentrated on the format of the product rather than on its intrinsic communication power. This example pinpoints exactly the way we've distorted this concept of creativity. Twenty percent of the qualifying score is for something we can't define and don't understand—an inordinate weight to assign to such a nebulous concept.

Is Creativity Essential?

Many people earnestly believe that creativity is essential to communication, but again they fail to define their concept of creativity. Most dictionary definitions of creativity are truly elusive. Creativity is uniformly defined as "the quality of being creative," which doesn't tell us much. Let's hear what two professionals contend. Ed Gray, a scriptwriter, discussed creativity in terms of art: "The very nature of the audiovisual [motion-media] business is art. . . . By utilizing the artistic skills inherent in an audiovisual production house, we are attempting to transfer information in the most effective way." But we're not sure what he means by "artistic skills," and note the word "attempting."[2]

David MacLeod, a motion-media producer, said, "In communications, creativity is the essence of packaging ideas and conveying them most effectively. It is not an add-on; it is not a peripheral. It's a foundation of our business, and it shines brightly through the lens . . . of projectors." His assertion seems to echo McLuhan's maxim, "The medium is the message."[3]

Accordingly, far too many designers and producers tend to define creativity in terms of production value—a dramatic storyline replete with actors speaking well-rehearsed lines on elaborate sets and in exotic locations, "clever" scenarios, digital effects, and lots of other "neat" stuff. Wonderful! Hope you have a great time. These folks expend resources on the razzle-dazzle, and cleverness (i.e., creativity) becomes the critical element that makes a "great" show. Did the show accomplish the communication goals set for it? What audience feedback do you have? Were the show's goals fitting, realistic, and worthwhile? Was the show worth the client's money, time, and other resources?

Here's a contrary opinion. My longtime friend (now deceased), Charles "Cap" Palmer, pioneer in information films, avers that "creativity is a vastly misunderstood and misused value, whose effectiveness in forwarding the sponsor's objective often is in inverse ratio to its noticeability. True creativity (effectiveness) is seldom tricky, bizarre, spectacular, or obviously 'clever'—in fact, it is more often basic and deliberately concealed" (emphasis added).[4] Palmer's philosophy established the basic tenet of just what creativity is in our information motion-media profession. If our shows are to be successful, we must embrace this philosophy and integrate it into our shows.

Shelton's Pronunciamento
Creativity must be basic and deliberately concealed.

Communication Fundamentals Often Ignored

What frequently warps the perspective of script designers and producers in the information motion-media profession is a hidden agenda that clouds their reasoning and jumbles their priorities. In essence, they've been bitten by the Hollywood bug and want to produce "movies," to create, and to be recognized and accepted by the film industry. They're "wannabes."

To illustrate how inane this mind-set can become, I'll cite one classic example. Some years ago, the U.S. Air Force produced a training film on the

KC-10 aircraft for Strategic Air Command crews. To grab the audience's attention, the opening scene shows the amply proportioned actress Bo Derek in a scene from the film *10* (released in 1979), wearing a form-fitting swimsuit and with her cornrow hairstyle, ambling along a beach. The scene fades into the outline of a KC-10 aircraft. Such a scene may well grab the audience's attention, but I'm not sure where. I spoke with an Air Force official who was responsible for this show. He told me that the justification for this opening sequence was, "We try to make it interesting . . . you can't just throw schlock at them."[5] Do you mean to say, sir, that a Bo-10 begets a KC-10?

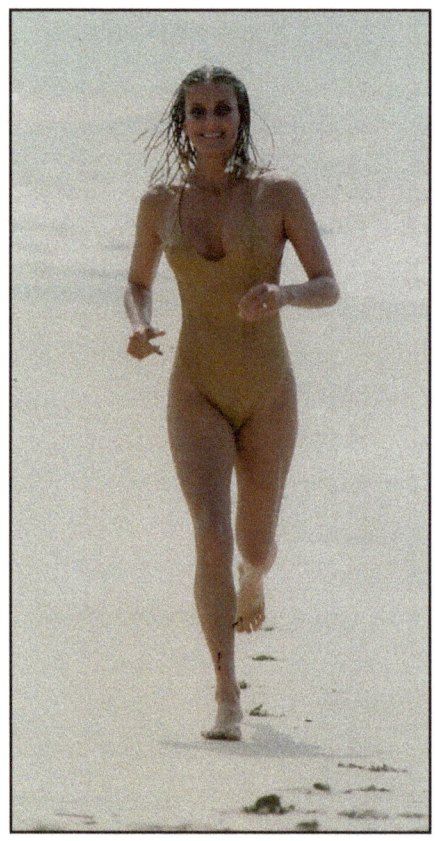

Figure 5. Ms. Bo Derek in the beach scene from Blake Edward's 1979 film *10*.
Photograph courtesy of Larry Edmunds.

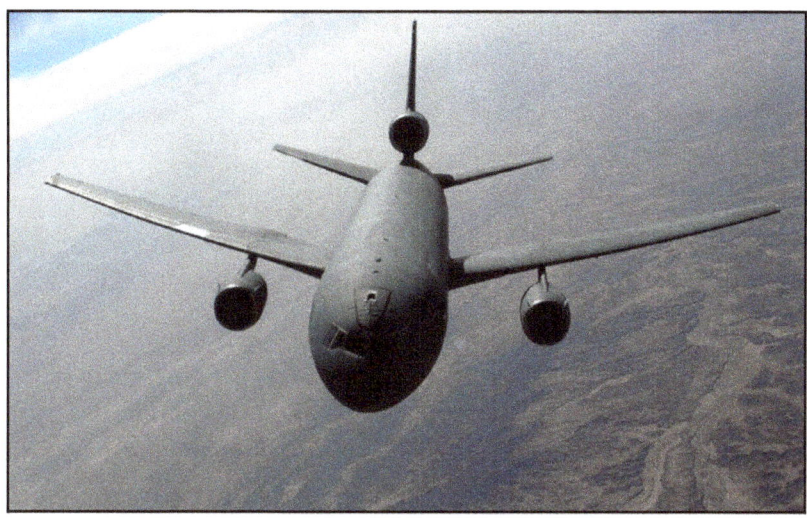

Figure 6. USAF McDonald Douglas KC-10 transport aircraft. Photograph courtesy of the U.S. Air Force.

Though this example may be at the extreme fringe of creativity gone awry, it's not atypical. I've seen all too many of this kind of gratuitous scene in shows over the years. It's the high-tech equivalent of third-grade refrigerator art. What's "interesting" is a show that motivates the target audience to action. It communicates.

Perhaps the most profound statement on creativity in motion-media comes from Jean O'Neal, producer/designer of multi-image shows. She has earned widespread recognition from her peers, winning a number of Best of Show and other topnotch awards. In a letter to me, she stated, "I still feel that it's much more difficult to do a one-projector show well than it is to do a 12-projector razzle-dazzle number. <u>You simply cannot cheat on a one-pro-jector show and pull it off</u>" (emphasis added).

Shelton's Pronunciamento
"Creativity," oftentimes, becomes discordant noise that hinders communication.

Interesting = Communication

No matter how uninteresting the information is to us, it will be interesting to the target audience if it is important to them and if they can realize benefit from the information.

John Grierson, the acclaimed documentary film pioneer, commented on these points many years ago when discussing the great influence his documentary films have had on a host of audiences throughout the British Empire (and the rest of the world). He said that "It is not the technical perfection of the film that matters, nor even the vanity of its maker, but <u>what happens to the public mind (emphasis added)</u>."[7]

Information motion-media shows should have an inherent dignity and a clear-cut purpose. Usually, a simple, straightforward presentation with voice-over narration, keeping the message in plain view, is appropriate for most situations. We use fundamentals of persuasion to lead the audience to a conclusion or decision that we've made obvious or inevitable. Extraneous gimmicks simply introduce noise into the communication process and should be avoided.[8]

Motion-Media Creativity

True creativity in our shows is a function of two factors that we control—communication value and technical achievement. We've used the appropriate grammar and syntax of the medium to integrate these factors into a totality that defines the creativity.

Communication value is a measure of the effectiveness of the motion-media show. To what extent are the stated objectives accomplished? My evaluation criteria are:

- Appropriateness. Considering the target audience and subject matter, the motion-media show is the appropriate communication tool to achieve the show's goals—**goals that are fitting, realistic, and worthwhile.**
- Information. The essential elements of information are developed logically, clearly, and succinctly, and are in the proper tone. The bulk of the information is encoded in the kinetic visuals. Audio elements reinforce the visuals, thus enhancing communication to full measure.

- Empathy. The show gains and holds the target audience's attention and involvement. It generates peak empathy. Communication is engendered to maximum effectiveness.
- Approach. The messages are couched in a tone and form that facilitates information flow and engenders empathy in our audience.
- Technical. The technical elements of the show ought to be appropriately skillful for the show's goals. (For additional information on technical details, see Chapter 5, page 44, Mascelli's *The Five C's of Cinematography*.)

Thus, I define a creative motion-media show as one that fulfills, to the maximum, the criteria that comprise communication value, and meets an acceptable threshold of technical achievement commensurate with the communication task at hand—and not much more.

We're professionals in this motion-media communication business. We achieve success not with the technology but with the creative use of our communication and technical skills. Such creative prowess is basic and deliberately concealed. Creativity, then, may well equal communication.

I want to conclude this chapter with quotes from two educational giants in our profession.

Doctor Charles F. Hoban Jr., (1906-1977), communication theorist, pioneer in communication research, and former Emeritus Professor of Communication at the University of Pennsylvania, casts creativity in its true perspective when discussing sophistication in the technology of film making. He maintains: "Hedonism is transparently triumphant. The gimmickry is great, but the message is lost or, worse, the wrong lesson is taught."[9]

Doctor Robert Davis, professor of communications at Florida State University and a visual scholar, notes that "Although technology is a wonderful thing, it is all too easy to become seduced by the hardware in audio-visual [motion-media] production, and forget the true goal of getting ideas across to the audience. True, you can dazzle them with special effects, but did anyone get the message?"[10]

Shelton's Pronunciamento

Keep to the fore the message and the audience.

Part II: Motion-Media

Chapter 5

Information Motion-Media

Communication objectives for our shows are extensive and diverse. They range from teaching psychomotor-skill development to cognitive learning, from attitude and behavior modification (work safely or stop smoking) to persuasion (buy Acme automobiles or vote for Samantha Smythe-Farthingale). Other objectives include correcting widespread misconceptions, improving morale, increasing awareness and sensitivity, informing, inspiring, instructing, introducing a new product or service, motivating, reporting on scientific progress, and training.

Information Flow in Linear Film/Video

Before we start discussing the grammar and syntax of linear motion-media, peruse Figure 7, the information flow in a typical motion media show. Start at "Message."

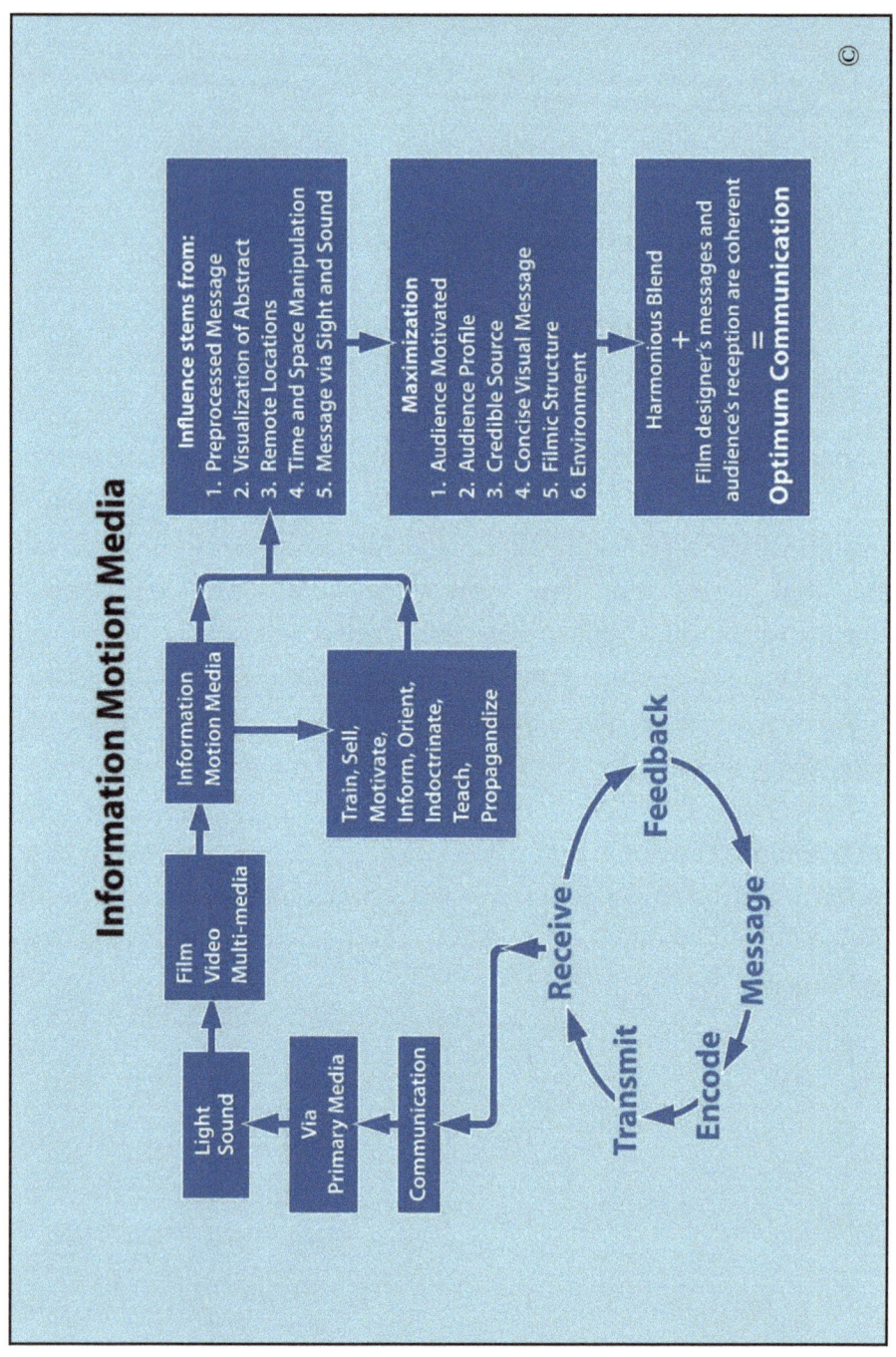

Figure 7. Information Motion-Media Communication Cycle.

General Discussion

As a mass-communication instrument, information motion-media has brought the world to the school-bound student and to all of us at our work and in our homes through television, the Internet, and social media of many stripes. Information motion-media is wonderfully effective in exploring the invisible, such as the electron flow in digital circuits; describing the immovable or unreachable, such as the moons of Saturn; reporting a current event—for example, the Nobel Prize awards; or selling an intangible product, such as group life insurance. Some of our shows are not quite so direct. For example, the Acme Computer Company sponsors a show whose goal is to teach senior citizens to use the Internet. An ancillary goal may be in a subliminal message designed to motivate the seniors to purchase Acme computers. A valid use of our media.

Unfortunately, we couch too many of our shows in a *mise en scène* that usually fails in its communication function. At first glance, they look wonderful, but there's rot inside. Following are several examples that I find particularly irksome.

All too often, I see information motion-media shows produced in a narrative filmic milieu where the message is buried in the entertainment. These shows may entertain the viewing audience and they may boost the script designer's and producer's egos, but more often than not, the entertainment elements engender noise that reduces the show's effectiveness.

Frequently we frame these shows in the hackneyed and patronizing style of the stilted scenario "old guy talking to the young guy" in all its many variations. See Chapter Seventeen, page 142, for details and the subsection titled "Stilted Scenarios" for my screed on this *mise en scène*.

Power of Information Motion-Media

Why is it that information motion-media is so powerful in communicating ideas? How are we using it to shape minds? John Grierson, founder of the documentary film movement in England, speaking of information and documentary films on the eve of World War II, said, "Like the radio and the newspaper, film is one of the keys to men's will, and information is as necessary a line of defense as the army."[1]

Grierson was one of the first to realize the *medium's* persuasive power—the overt and subliminal intrusion of ideas and emotions into a person's psyche. In 1935, he said, "Cinema is neither an art nor an entertainment, it is a form of publication, and may be published in a hundred different ways for a hundred different audiences. . . . [However,] the most important field is propaganda."[2] If we define propaganda to mean marketing, his perceptions were uncanny.

A classic example of the propaganda film is William Wyler's Academy-Award-winning documentary *The Memphis Belle*. This film depicts the flight of the Boeing B-17 bomber dubbed "Memphis Belle" on its 25th mission over Germany. This time, the nine-man crew's target is the submarine pens at Wilhelmshaven, Germany. I would suggest that *The Memphis Belle* is the greatest film from World War II. The crew's actual combat scenes are thrilling, chilling, and compelling. A must see for all—a memorial to the sacrifices of the airmen of the 8th Air Force, the "Greatest Generation."

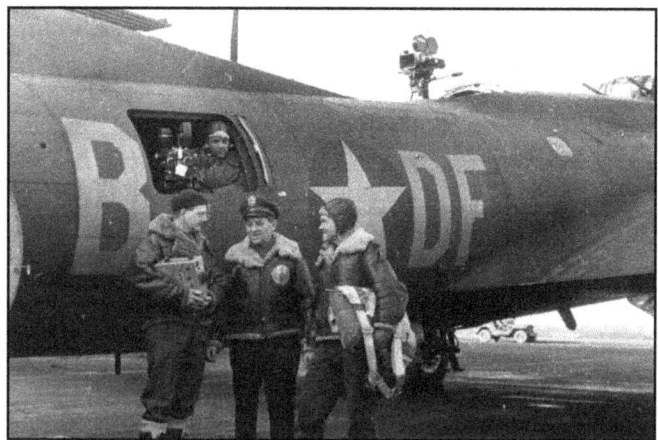

Figure 8. Lieutenant Colonel William Wyler directing a scene in his Academy-Award-winning film *The Memphis Belle*. Wyler is at the Army Air Corp Base in Archbury, England, 1943. He is the officer in the center of the photograph wearing the flat hat. Photograph courtesy of Mrs. William Wyler and the UCLA Film Archives.

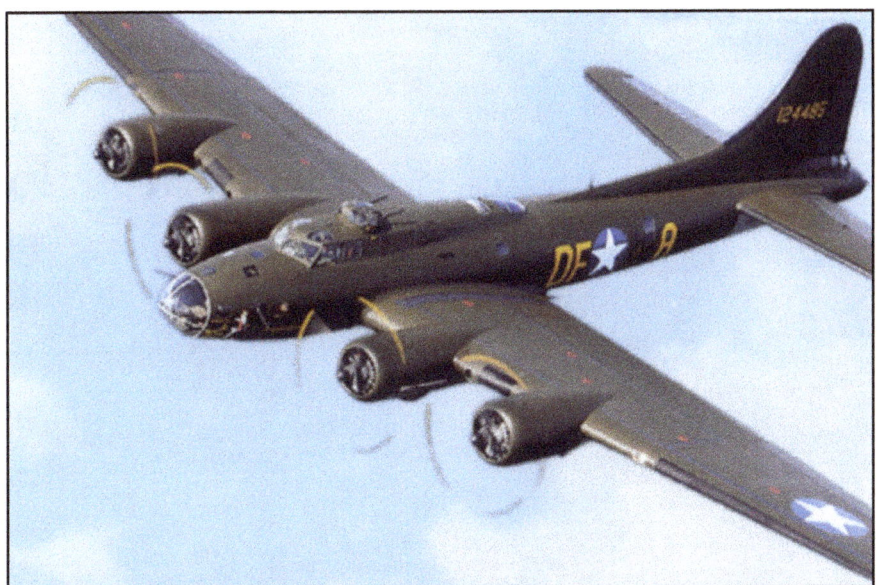

Figure 9. The Boeing B17F dubbed "The Memphis Belle" in flight.
Photograph Courtesy of the U.S. Air Force.

Leni Riefenstahl's *Triumph des Willens* is another outstanding example of a propaganda film that generated intense empathy and resolve in the target audience—the people of Germany, languishing under the largely French-imposed, wrathful terms of the Treaty of Versailles (1919). This powerful film chronicles the National Socialist German Workers' Party (Nazi) Congress in Nuremberg in 1934. There, Adolph Hitler and other party functionaries speak to rally the *volk* to their overarching theme: Germany is to return to being a great power in Europe. The exceptionally well-crafted montages of the massed Sturmabteilung (Storm Detachment, SA), *Schutzstaffel* (Protection Squadron, SS) troops is classic and worthy of intense study by all in this profession.

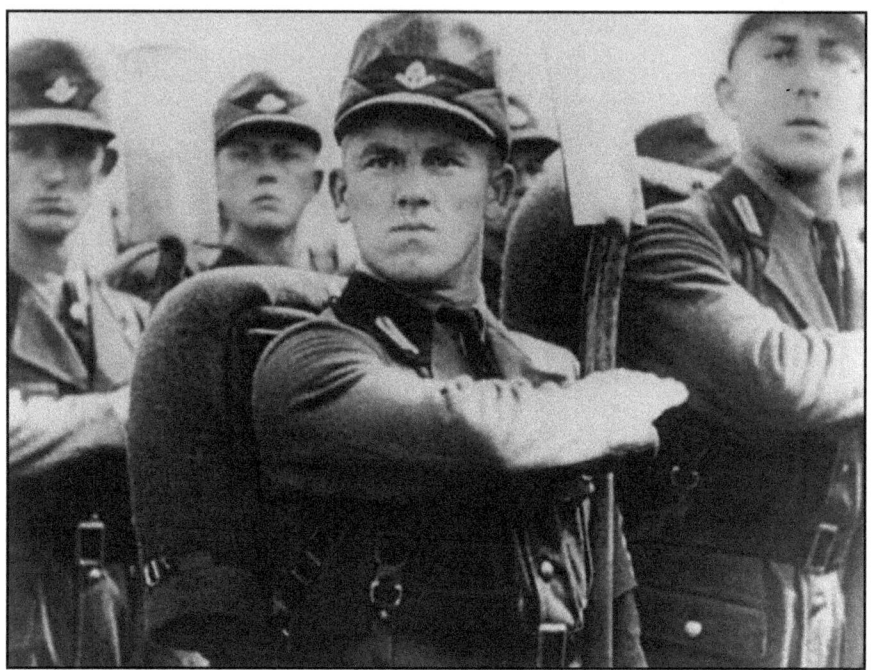

Figure 10. A frame of Riefenstahl film *Triumph des Willens* featuring "soldiers" of the Sturmabteilung (SA). Photograph courtesy of the National Audiovisual Center.

Background

In the early days, information film was called sponsored film because industrial organizations first used it as a marketing tool. According to the curator of Eastman House in Rochester, New York, sponsored films were produced shortly after the invention of motion pictures, the first appearing in the early 1890s. Film historian and information-film producer Walter Klein said in his book *The Sponsored Film* that the Standard Oil Company of Indiana sponsored a film in the early 1900s. General Electric produced one in 1907, National Cash Register in 1911, and the International Harvester Company sponsored *Back to the Old Farm* in 1911. By 1930, there were only about thirty films that could be called "educational."[3]

During World War II, the use of information film exploded as a powerful communication tool to train our military. Significantly, it was our archenemy, Field Marshall Wilhelm Keitel (1882-1946), chief of the High Command of the German Armed Forces, who commented prophetically that (paraphrased) Germany could not win the war because of the way Americans used film to train their soldiers.

Factors Affecting Influence

We can identify five major factors that apply to all information motion-media:

Motivation. The audience must be motivated in order for communication to occur. Each one has to be convinced that it is to his or her advantage to be in the viewing environment and to pay attention.

Credibility. If a persuasive information motion-media show is to be successful, the audience must hold the film designer, sponsor, and the medium itself in high esteem in terms of credibility and prestige.

Audience Profile. Audiences come in all sizes and compositions. Individual temperament, background, understanding, and values are singularly diverse. If we are to communicate effectively with the target audience, we must know who they are. That is, we need to know the audience's composition—the audience profile. Optimum communication occurs when we couch the messages in a tone and a *mise en scène* that are familiar, relevant, and sympathetic to the audience. Such messages engender understanding, acceptance, involvement, and empathy. And without empathy, there's little or no communication.

Content. There are a host of content factors that apply. I've found the following general points to be crucial for most of our shows (I recognize that exceptions exist for special applications):

- The messages need be short and to the point. Too many details cause a mental "short circuit" in our audience.
- Cute gimmicks often backfire and become noise that obscures the primary message.
- Most messages should be visual. About 70 to 80 percent of the information should be in the pictorial content.

Auditory messages must be secondary and used sparingly only to reinforce the visuals.

Structure. How the messages unfold is critically important. The following list contains ideas on filmic structure that I've found particularly useful:

- Anticipation heightens audience awareness for a coming event.
- A "heads-up" alerts the audience to the incoming message.
- Requirements tell the audience what is expected from them after the viewing, and explain why.
- Repetition is more than simple iteration; it is the varied presentation of the major points, concepts, and meanings.
- The optimum development rate allows the audience time to assimilate the new information or concepts and allows time for reflection.
- Participation involves overt sharing in the communication process, exemplified by interactive multimedia.
- In shows with sociological themes, a narrative (dramatic) structure is appropriate.
- Summation reviews key points and presents a cohesive survey of their relevance.

Cinematic Techniques

In addition to the influence factors that we've discussed so far, we need to explore the various cinematic techniques that influence the structure of our shows. Such techniques are detailed masterfully in Joseph V. Mascelli's (American Society of Cinematographers) landmark book, *The Five C's of Cinematography*.[4] I urge you to study Mascelli's book in detail. Here is my summary of Mascelli's "Five C's":

1. Camera Angle. Determines the audience's viewpoint, engenders emotion, and sets the scope and perspective of the scheme. The basic camera angles are extreme long shot, long shot, medium shot, two shot, close-up, and other specialized angles. Additionally, Mascelli classifies shots as:

- Objective. Audience sees the action through the eyes of an unseen observer, as if eavesdropping. This is the basic cinematic shot.
- Subjective. (Sometimes dubbed zero-camera-angle.) Subjective camera is that camera angle which shows the action from the eye position of the audience. Audience sees the scenes as if they are see-

ing them through their own eyes. This angle is preferred in detailing procedures in close-up shots.

- Point of view. Audience sees the scene from the point of view of an off-screen player. Puts the audience cheek-to-cheek with this off-screen player.

2. Close-Up. Eliminates all nonessentials in the scene and focuses on the critical element of information. Engenders dramatic impact and visual clarity, involves audience in details of the scene. (Even though close-up usually is included in camera angles, Mascelli treats close-up as one of the Five C elements.)

3. Composition. Arranges all pictorial elements in a shot with the goal of forming a unified, harmonious milieu that forces the eye to the center of interest and stimulates the audience to the "appropriate" mood. Lighting is a prime factor in setting mood.

4. Continuity. Makes a smooth, logical flow of visual imagery from shot to shot and from scene to scene. Establishes coherence; defines the path in the plastic medium. There are exceptions for special considerations. Nowadays, discontinuous cutting between scenes is in vogue. Sometimes, sound continuity is the thread of coherence.

5. Cutting (editing). Editing is the function that in large measure manipulates the plasticity of the medium. Editing determines the measure of coherence and tells the "story." Editing is largely responsible for capturing and holding audience interest through scene selection, sequencing, tempo, screen direction, interplay of scenes, and use of effects. Superior editing is essential to a successful motion-media show.

Mascelli's "Five C's" define the basic tenets of filmic design—the grammar and syntax of motion-media. It's with filmic design that we script the kinetic visuals and produce our shows.

If We Can Do It in Five, Why Take Ten?

Most linear information motion-media shows are too long, saturated with too much information and too many concepts. I've seen far too many shows that ramble incessantly, pounding the audience into catatonia with a barrage of overwhelming and gratuitous trivia that obscure the essential points and significantly reduce communication. In essence, too many messages confuse and bore the audience.

Our information motion-media shows are competing with a host of other interests and distractions for the valued time and minds of our audiences, yet we continually ignore this fundamental fact. We make our products too long and ponderous. We pack them with too much information—information that usually is beyond the comprehension, concerns, and retention of our audiences.

Shelton's Pronunciamento
The shorter the better.

I'm convinced that most topics with which we deal ought to be developed quickly. For instance, the six-minute, 1973 motivational film *The Man From LOX,* with its sharply focused communication thrust, was eminently successful and became a standard in its field for many years.

Several researchers have conducted empirical investigations into the problem of communication overload. In the aggregate, their conclusions are much the same: as more and more information is presented, interference (noise) is induced in our audience, resulting in reduced empathy and thus less-efficient communication. It's an inverse proportional ratio: as we increase the information content, there's a proportional decrease in the amount of communication accomplished.[5] Most audiences can comprehend only a discrete amount of information in a viewing because of the linearity of film and video.

The liability of linearity argues well for short motion-media shows—five to seven minutes—designed with a rapier-like communication thrust aimed directly at the target audience. In such shows, there's no tedium, preaching, or ennui.

Today, we're dealing with sophisticated audiences who are experienced in receiving messages in thirty- and fifteen-second blasts (e.g., television commercials). Thus, in today's communications environment, successful information motion-media shows need to transmit information in highly concentrated, relatively small doses to select, well-defined audiences. All other factors being equal, such shows gain the communication advantage by:

- Using a memory aid to reinforce what the audience already knows about the topic
- Highlighting three or four key points of one central theme
- Setting these points in relevant perspective in terms of background and related facts or concepts
- Tying the points together in a related chain
- Reinforcing critical points
- Getting "The End" title on quickly

The memory aid, in particular, is an excellent device to save lots of screen time. It enables us to get our show off to a running start and to keep it at a fast pace.

Television Commercials

I consider the television commercial to be the definitive information motion-media show—its goals, on the whole, are to buy, vote, or contribute—hawking everything from army recruitment to zinfandel wine. Producers of television commercials, forced by the constraint of getting their message across in just a few seconds and driven by intense pressure to succeed, pioneered a metamorphosis in our profession. Using linearity to advantage, they developed a neo-Orwellian communication tool, engineered with the latest technology, populated with glamorous people, and laced with seductive messages. The messages are singular in purpose, rapid-fire, repetitive, and largely inescapable. The television commercial molds minds and shapes lifestyles effectively, quickly, and categorically. There are four fundamental psychological keys to television commercials:

- Send short messages to realize effective communication.
- Blur the boundary between reality and fantasy.
- Stimulate the self-gratification libido.
- Reinforce impulsiveness and selfishness.

Though it may be awkward to admit, we need to realize that these same four principles apply to the entire genre of information motion-media, albeit perhaps not always so flagrantly or intensely. Nonetheless, they are fundamental to our profession. No matter how we rationalize or what we call it, our task is to manipulate the minds of our audiences. We influence the behavior and thoughts of others, whether they want to be influenced or not.

In essence, the television commercial is a single-concept information film that has a deep psychological punch. Of utmost significance, it's in the television commercial that many new, often radical, production concepts and techniques are used first. From my perspective, the television commercial is the research, development, test, and evaluation function of our profession. As such, there is much we can learn from these highly effective, single-concept communication tools.

Unfortunately, far too many television commercials are overdone—lots of razzmatazz, but there's rot inside. On viewing such a commercial, the viewer may ask, "Beautifully done. (*Pause.*) I wonder what it is all about?" Particularly irksome are those pharmaceutical commercials that hawk prescription medicines. They open usually with a recovered patient telling us how wonderful the Acme Alchemy nostrum is. Fair enough. Then the all-knowing narrator spews all the evil side effects that this prescription drug can inflict on the patient. Concurrently, a series of irrelevant visuals flash across the screen. Egads! This scenario is "bass ackward." There's no coherence between visuals and aural, and the message is in the narration and not in the kinetic visuals, where it ought to be.

The Prophecy

Fortunately, some in our profession understood and heeded the power. Taking his cue from the early television commercials, Cap Palmer, a missionary for the short, single-concept information film, prophetically said, "The educational film we have known is doomed . . . tomorrow's audiovisual will be single concept, from thirty seconds to a few minutes long."[7] The important point is that Palmer's prophecy of doom was based on the general ineffectiveness of the traditional, long-winded, overbearing training films of the day. He saw the compelling need for the short, single-concept film to break through the traditional barriers of cost and clutter and open up new worlds of effective communication.

Short, Single-Concept Motion-Media

The short, single-concept show is in ascendancy—the television commercial, for instance. It is becoming the most critically important communication tool we have, a tool that readily engenders communication. Messages in concentrated bursts are the intrinsic elixir of the short, single-concept, linear information motion-media show. They are inherently simple, direct, cogent, and succinct. In other words, every second counts. Every frame counts. The audience instantly engages. The show gains momentum all the way to the end credit. The audience should leave a little wiser, a little sadder or happier, a little different.[8]

Linearity of the medium augurs well for short films/videos—five to seven minutes, and with pinpoint communication thrust. I've found this type of show to be significantly more effective than those that laboriously continue on, and on, and on.

Not a Panacea

The short, single-concept information motion-media show isn't the ultimate answer to all persuasive communication tasks. Nothing is. In some instances, the topics, goals, and audiences are not suited to this type of presentation. To get our communication task done right, we need to consider the full repertoire of communication alternatives—different forms of motion-media, or some other media. Or, perhaps some combination of media is appropriate. I've found that combining media is exceedingly successful in communicating many types of complex messages. One excellent way to reinforce a show's communications is to have available well-scripted and illustrated information, either printed or on the Internet.

Chapter 6

Linear Media

As communication media, there is no difference between film and video. Both develop information linearly, one point at a time in sequence. The principal differences are technical: the way information is captured, stored, manipulated, and distributed: either chemically or electronically. Nowadays, even this difference is disappearing. For instance, a show may be shot on film, edited in video, and distributed in film, video, or a variety of digital formats.

Frequently, I've found that there is no one "best" communication medium for the problem at hand. Linear film/video is particularly useful for imparting broad-based concepts to inform, motivate, sell, orient, heighten awareness, and other "soft" goals. A simple, straightforward approach is usually best for communicating these sorts of goals.

The film/video media is not effective usually for communications that require detailed, long-term retention of specific facts or procedures. For instance, perceptual motor-skill training. Instead, interactive multimedia is the medium of choice for such long-term learning.

Sometimes, linear film/video is effective in teaching. Such a show is short, has no more than four or five key points, and is narrowly focused. If a film/video with cognitive goals does not exceed these criteria, it's almost always ineffective. Otherwise, our audience just can't remember all the details, gets bored, and loses interest. For example, a few years ago I critiqued a film/video that was intended to teach sales personnel about the products that the sponsor manufactured and sold. The show was fifty-eight minutes long and had about fifty key points. The ultimate blow to communication was that the show's *mise en scène* was the classic, stilted scenario of the "old guy talking to the young guy." In this case, it was the store manager (male) talking to the sales associate (female)—both facing the camera. This super-expensive show was a total failure. Admittedly, some of the audience might remember a few of the key points, but how could they put these points in context? I suspect that boredom and resentment in the audience negated whatever minute positive communication value the show had.

Unfortunately, film/video's tinsel glamour seduces too many of us and we end up producing "movies" rather than communication products. We couch our messages in entertainment. I've seen far too many of these sorts of shows. They always fail. They're too complicated, long, expensive, and tedious. Wake up the audience when the lights come on!

Advantages of Film/Video

Film and video offer a host of advantages as communication media. Conversely, they have serious disadvantages. Let's focus first on the advantages. I've listed several in descending order of importance. Also, in this section we will begin to explore the grammar and syntax of the media—the tools of filmic design.

Visual Medium. Film/video's most important characteristic is that it's a kinetic visual medium. Each photographic scene is a reproduction of the images and movements of events (real or simulated). Of course, we can distort the scene with digital manipulation. The power of film/video as a communication tool is illustrated by the old Chinese proverb "a picture is worth ten thousand words." A study conducted by the National Audiovisual Association a few years ago found that in a motion-media show 83 percent of learning comes from the visuals. Audiences remember *long-term* 30 percent of what they see, and 50 percent of what they see and hear simultaneously.[1]

Shelton's Pronunciamento
Motion media is a kinetic-visual medium.

These remarkably high learning and retention figures result from an audience's ability to assimilate visual information at an extraordinarily rapid rate. Dr. Bryan Wilson Key of the University of Nevada, Reno (author of *Subliminal Seduction*) conducted studies that found that, on average, a person can grasp, process, and store visual information in as little as 1/1,000 of a second.[2] Also, the more familiar the audience is with the visuals, the more there is a non-linear increase in assimilation and understanding. Key indicates that all our sensory equipment operates simultaneously and continuously, and that the upper limit of our reception ability is, as yet, unknown.

Thus, as a kinetic-visual medium and with the camera in deft hands, film/video's power is projected exponentially. Camera distance and angles change visual perspective. Long shots give orientation and perspective to the scene. Medium shots show comparisons. Close-ups give detailed examination. Deft use of the camera attracts the audience's attention to objects and movements that are only marginally perceptible in a scene. Transference of attention occurs in a scene by a movement (perhaps ever so slight), a light shift, a change in composition or perspective.

Multisensory. The primacy of film/video is that it's multisensory. Sight and hearing are stimulated in concert. This dual medium quality, with the right mix of sight and sound, has a powerful synergistic effect on audience retention of our messages. Such a "right mix" of sight and sound portends maximum opportunity for optimizing communications and engendering empathy in our audiences.

What's the right sight/sound mix? We find a working answer in a wide-ranging series of psychological studies conducted by Charles F. Hoban Jr. and Edward Van Ormer, communication scholars at Pennsylvania State University, in the 1950s. Their studies investigated the power of films to communicate ideas. One critical finding was that about 70 to 80 percent of the messages should be encoded in kinetic visuals—such filmic technique is the key element in successfully transmitting information. Accordingly, aural information, commentary, a minor contributor, should not exceed 30 to 20 percent of the total information content of a show.[10] I'll state this ratio as 75/25 percent.

Shelton's Pronunciamento
Encode 75% of the messages in the kinetic visuals. Limit commentary to 25%.

And, the auditory information, narration or dialogue, must not introduce communication noise. That is, the spoken word, mostly voice-over narration, must have absolute coherence with the visuals. These two cinematic elements must show and tell identical messages; otherwise, the mixed messages become noise and confuse our audience. We must avoid that old saw found too often in scripts that describes the kinetic-visual information as "a variety of scenes to complement the narration."

For example, I've seen many shows in which serious noise emanates from a narration that's an incessant harangue of drivel or a *mélange* of technical trivia that dulls the senses—written as if the audience would read it rather than hear it. Attempting to comprehend what is being <u>said</u>, audiences focus on the commentary. Meanwhile, the kinetic visuals, mostly irrelevant, stream onward and onward. The audience gets mixed messages. Eventually, they'll give up by escaping mentally or physically to a more comfortable environment. Such a show does not communicate much.

Shelton's Pronunciamento
Commentary must be used only to explain or amplify what the audience cannot perceive from the kinetic visuals yet must know for a complete understanding.

For instance, in my award-winning ten-minute show *299 Foxtrot* there's minimal narration, perhaps only a minute's worth if it were spoken continuously. This film depicts the resurrection of a Boeing B-29 from an aircraft boneyard and its flyaway to a museum in Topeka.

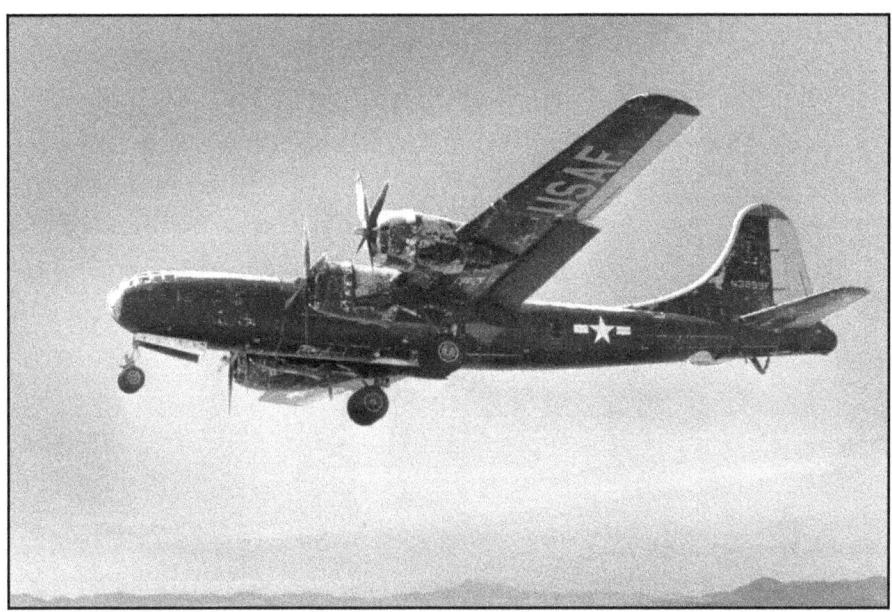

Figure 11. Boeing B-29 dubbed *299 Foxtrot* shortly after takeoff from Armitage Field at the Naval Weapons Center, China Lake, California. Photograph from the author's private collection.

In motivational and attitude-modification shows that deal with sociological themes, such as drug addiction, alcoholism, child abuse, and sexual harassment, we couch the key points in a dramatic narrative. Professional actors in authentic locations, wardrobe, and props, foster a realistic scenario. Such shows, if designed and produced professionally, can engender intense audience identification with the characters, setting, and theme to create enduring empathy.

The pace of the narrative shows tends to be slower and longer than the voice-over narration type because we've couched some of the key points in the actors' dialogue—aural communication.

Let's explore one example. The show is *Gambling Addiction and the Family*. (See Appendix Four, page 228 for a segment from the show's script.) The objective was to communicate the destructive effect the wife's gambling addiction had on her family. The scene begins in the family living room. It is the final confrontation between the husband and his chronic-gambler wife, who abuses the children and is destroying the marriage.

The audience sees and hears the anger and frustration of the husband and the craven denial of the wife. Note that the kinetic visuals are minimal. The vast majority of the information is in the dialogue. However, facial expressions, body language, and staging are contributing factors.

Noted filmmaker and cinema historian, the late Karel Reisz of the British Film Academy seems to offer another perspective. In discussing the robbery sequence in Carol Reed's 1946 film *Odd Man Out,* he wrote, "The dialogue track does not anchor the visuals by conveying important information, but adds to the total effect on a contributory rather than a primary level."[4]

Reisz's comment is valid for narrative films produced to earn a profit by entertaining paying audiences—films often suffused with dynamic kinetic visuals. Our shows are produced to influence the minds of our target audiences, not to entertain them.

Contributory aural elements

Music, sound effects, and background prattle set mood, add fictional reality, and enhance the pace established by the kinetic visuals. These aural elements must be <u>natural</u> and <u>unobtrusive</u> in the scene to the degree that the audience does not "<u>hear</u>" them. Interference, or noise, is established quickly when these aural elements become preeminent in volume or importance, or when they are incongruous. A professional sound mix is essential to blend these sound elements into a coherent whole that reinforces the kinetic visuals.

Shelton's Pronunciamento

The audience ought not hear the sounds in motion media.

Sometimes, music or dance is the major theme in film/video. Such shows fall into the enrichment genre and are not strictly informational. Two examples are Norman McLaren's *Ballet Adagio* (Bronze Plaque, Columbus Film Festival) and Allan Miller and William Fertik's *Bolero* (Academy Award).

Mass Media. A major advantage of linear media is that the same message can be transmitted to mass audiences in a myriad of locations—and, if need be, in real time. The messages are stored in physical and electronic media for audience preference viewing. With the flexibility of the current distribution media, a person can review the messages multiple times for reinforced communication.

Visually Sophisticated Audiences. As a result of many years of incessant bombardment by the mass media, today's audiences are attuned to receiving copious amounts of kinetic sight-and-sound information quickly. Accordingly, our shows must be sharply focused, intensely relevant, and intensely empathetic.

Empathy. Expertly produced shows engender empathy in the target audience. The more enduring the empathy, the more effective the communication. That's our challenge. In the preproduction phase of a show, we develop its filmic design to entice the audience to identify with the *milieu*: location, characters, setting—all the elements in our show. For instance, the messages must be

- relevant
- couched in familiar terms and settings
- believable—(no matter how false they may be)
- congenial—having an asking, sharing, and involving tone

The intensity of audiences' empathy is a direct function of their *willing suspension of disbelief.* The folks inherently know that the images they are seeing are fictitious and are prone to disbelieve them. However, as the empathy increases, the audience becomes emotionally involved and temporarily accepts the fiction—they suspend their disbelief.

Shelton's Pronunciamento
Audiences willingly suspend disbelief in carefully targeted shows.

Plastic Medium. We distort reality with editing, special (digital) effects, and camera techniques. For instance, we

- manipulate time and space. In post-production we insert a close-up (CU) shot into a master scene to either expand or compress time.
  - An editorial cut changes the scene from outer space to the

interior of the Oval Office.

- High-speed or slow-motion photography distorts familiar forms of motion.
- visit any location nearly instantaneously and make remote locations congruent with our theme.
  - view the outer reaches of space or see the inner working of the particles in an atom of oxygen.
- use special effects (digital) to warp familiar forms into a potpourri of unreal kinetic visuals.

By juxtaposition of scenes and sound, we combine individual and unrelated elements into a new relationship that they do not inherently have, thus creating dramatic, powerful psychological effects: "to explode ideas in the heads of our audiences," as John Grierson said.[5] These unique capabilities result from editing shots into a cinematic rhythm that manipulates relationships in an infinity of unnatural combinations and associations. Emotions and ideas are juggled and interpreted in unusual ways that evoke audience responses to different perspectives, which may be real, distorted, or false.

Sergei Eisenstein noted that "two film pieces of any kind, placed together, inevitably combine into a new concept, a new quality, arising out of that juxtaposition."[6] For example, if I were to edit a three-shot sequence consisting of a close-up (CU) of a computer screen (photographed in Baltimore), followed by a close-up of the hands of a person working a keyboard or mouse (photographed in Dallas), followed by an extreme close-up (ECU) showing changes on a similar monitor in Helena—assuming backgrounds and lighting are similar—our audience would conclude that the hands are causing the changes on the monitor.

Also, scenes that are dissimilar, unrelated, or non-sequential can be joined into a meaningful relationship by juxtaposition and adroit screen-direction manipulation. This technique is prevalent in historical documentaries composed of stock and news footage: *Victory at Sea* is an excellent example.

We use time and space manipulation for detailed study of fast-happening events, for scientific analysis, for understanding natural phenomena, or simply for artistic purposes. Time-lapse photography compresses into a few seconds the hours or days that some events take to complete—cumulus clouds building or flowers blooming. Conversely, high-speed photography expands near-instantaneous events into seconds, minutes, or hours: for example, an explosion or a staged automobile crash shown in slow-motion or the sack of a football quarterback.

Shelton's Pronunciamento

Motion media's paramount characteristic is manipulation of time and space—the plasticity of the medium.

Designer Selection. We show the audience precisely what we want them to see, and when we want them to see it. With scene selection, we have total visual and audio control. In a live stage play, the audience can choose by concentration and selective viewing what they see, be it the entire scope of the stage or just the face of an actor speaking. In film/video, the audience can see only what we show them. Recognize that the audience has some selective viewing decisions in certain types of scenes, but this is severely limited in terms of the total viewing experience in motion media.

Let's explore the sequencing of shot types to develop a linear scene.

- Long shot (LS). Establishes "This is the place." Sets the ambience, the players, the "things" that interact in the scene, the mood, and the background.
- Medium shot (MS). Reinforces the settings and focuses on the key players or object(s) that are the center of attention.
- Close up (CU), and extreme close-up (ECU). Determine, in large measure, what is important—who it is, how it works, who is doing what to whom or it. The more full-frame an action is, the more emphasis it has. It's in the close-up and the extreme close-up, and to some extent the medium shot, that the essential elements of information (key points) are successfully transmitted. Additionally, a series of CUs or ECUs edited into a montage scene significantly en-

hance the importance of an action.

- For excellent montage examples, view Eisenstein's "Odessa Steps" scene in *The Battleship Potemkin,* and Riefenstahl's *Sturmabteilung* troops scene in *Triumph des Willen*—both films noted in Appendix Two, pages 194 and 202.

The sequencing of shots defines the message's relevance. For example, consider this fundamental filmic-design sequence: LS, MS, CU, ECU, CU, CU. All shots depict related and sequential action. When the essential elements of information are contained primarily in the CUs and the ECU, the audience should immediately comprehend the information's meaning and importance. And they'll understand its relationship to the preceding background information established in the LS and MS.

Any scene should be on the screen/monitor just as long as the audience continues to receive new and important information. If the scene dallies, our audience's eye exhaustion point is reached quickly. The folks tend to get bored and to mentally focus on something else.

Abstract Visualization. Film/video is an excellent medium for visualizing abstract concepts. Through a short animation sequence we could make understandable the economic concept of gold in international finance or the effect of oil on the world's economy for tenth-graders.

Remote Location. Film/video takes the audience to realms far removed from the screening/viewing site. Our cameras go anywhere on the earth's surface, under the oceans, and into space. This advantage is not limited to geographical locations; it encompasses the total range of possibilities, including microphotography of bacteria, microphotography of molecular structures, or astrophotography of a distant nebula. With fiber-optic photography, we can see the aortic valve functioning. Through animation, we visualize something that is inaccessible by reason of size or non-availability. For instance, we can explore the inner workings of the human circulatory system or look inside a transistor to see the electrons flow. With specialized cameras, we record microwave, infrared, and radar images.

Tailor-Made. Every show is unique. We design each show to resolve a particular communication problem. The precision-communication thrust of the show is concentrated on the essence of the problem for the specific target audience.

Silver Screen Magic. All factors considered, I've found that audiences generally tend to be receptive to messages contained in information shows. Such acceptance stems from some sort of "silver-screen magic" that I cannot identify. This magic comes from an arcane and inherent authority attendant on all well-produced shows. Simply said, it springs from the fact that if the show has been made, "it must be important." I've found that our audiences usually are receptive to an upcoming show and bring lots of goodwill and anticipation. They're willing to be open and follow our development as long as we engender empathy and maintain it—their willing suspension of disbelief.

These advantages of film/video I've noted make motion-media a powerful tool in solving many communication problems. It's not infallible, however.

Disadvantages of Film/Video

Information film/video has disadvantages that make it unsuitable for some communication scenarios and tenuous for others. Let's look at the disadvantages of linear media. (I discuss multimedia in Chapter, 9, page 87.) I've listed them in descending order of importance.

Linearity. The prime disadvantage of traditional film/video is that it's a time-based medium. The show is developed sequentially and, in the usual venue, audiences view it sequentially. They have no time for reflective thought or for review of difficult sequences.

Transitory. In the traditional mode, information viewed on film/video is evanescent. That is, the instant the audience has viewed a frame, the image is gone. In linear viewing, detailed information is lost so quickly that, if it's not reinforced, it has only marginal validity. Usually, after a few days, our audience remembers only a few impressive highlights. A handout, such as a liberally illustrated, printed-word document, is an excellent reinforcement medium.

Today, we can store our shows digitally on portable electronic devices. Accordingly, our audiences have the option to view a linear film/video as they would a multimedia show; that is, to peruse the information randomly—review, freeze-frame, skip, go forward or back at multiple frame rates, and so forth.

Alien Perspective. We encode the information in a motion-media show in a *mise en scène* that reflects our perspective to engender audience empathy. Such a perspective may well be alien to some in our audiences. Simply, alien perspective adds noise into the communication process. Careful preproduction audience analysis will alleviate some of this problem. But we can never completely eliminate it because of the uniqueness of each individual in our audience. The easiest show we can produce is designed for one person. Our audience analysis enables us to draw an accurate psychological profile of this individual. Accordingly, we tailor the show in the precise terms of this profile to optimize communication.

Other problems arise for those who cannot or do not want to follow the development we've set, as to what's seen, how it's seen, and where it's seen. For some, the pace of the show may be either too fast or too slow for their individual comprehension rates. (Multimedia is not encumbered by these properties.)

No Room for Error. On the whole, motion-media commands attention. Therefore, one absolute guarantee of failure in a show is to be dishonest or inaccurate in any way. Today's audiences are too sophisticated to be deceived. A show's total credibility is lost quickly if the audience detects any misstatement of fact or visual distortion that's intended to deceive.

Within my experience, I've found that one way to have a visual or aural distortion accepted is to tell the audience unabashedly that the scene is a fake, for whatever reason. For example, in a show intended for student aviators, our narrator might say, "We are using a model airplane because it's easier to control in this demonstration of approach procedure." The audience usually will accept such a truth stated forthrightly and empathize with the scene.

Conversely, deception is the norm in propaganda shows. Such deception usually is not overt, it's subliminal. Expertly done, it's ingratiating, effective, and dangerous. I would suggest that Leni Riefenstahl's 1935 film *Triumph des Willen* is a prime example and perhaps the all-time classic. (See Figure 2, page 11, and Figure 10, page 42.)

Shelton's Pronunciamento

Triumph des Willen propelled the German *volk* to fully support the Third Reich.

Chapter 7

The False Reality of Motion-Media

There is no reality in motion media—there's only the illusion of reality. In empathetic shows, our audience willingly suspends disbelief in what they see and hear.

Theories of Film Reality

There are several theories that explore the concept of reality in film. Of these, I've chosen five that have historical and philosophical perspectives for our discussion in this chapter.

Part-Whole Theory. As expounded by Sergei Eisenstein and V. I. Pudovkin, this theory contends that individual scenes in themselves are un-filmic and only fragments of reality. Only by editing scenes into coherent montage sequences does film become art, and reality emerge. The Odessa Steps montage in Sergei Eisenstein's 1925 *The Battleship Potemkin* is a classic example.[1]

Relation to the Real Theory. This theory, developed by André Bazin and Siegfried Kracauer, rejects the idea that film art and reality result from creative editing. Rather, Bazin asserts that if individual scenes are faithful reproductions of reality, only a simple assembly of scenes is needed to make the total film art and an accurate replication of reality. He notes that "we are forced to accept as real the existence of the objects reproduced, actually re-presented, set before us [in film], that is to say, in time and space."[2] Walter Ruttman's 1927 film *Berlin: Die Sinfonie der Grossstadt* (*Berlin: Symphony of a Great City*) is an example of Bazin's notion.

Kino-Eye Theory. This theory, propounded by Dziga Vertov, asserts that the camera, as an extension of the human eye, has the ability to penetrate every detail of contemporary life.[3] Essential to this theory is the concept that the film designer must never stage a scene or interfere with a spontaneous action among people or in nature. The film designer's only job is to record events as they happen naturally. Through use of technical devices and montage editing, film evolves as a mirror of reality Barbara Kopel's 1975 *Harlan County, USA,* tends towards this goal.

Creative Actuality. John Grierson, founder of the documentary film tradition, defined documentary film as "the creative treatment of actuality." Grierson illustrated his concept by describing the film *Moana: A Romance of the Golden Age* as "... a visual account of events in the daily life of a Polynesian youth, [it] has documentary value." He continued, "it became an absolute principle that the story must be taken from the location, and that it should be the essential story of the location."[4] An excellent example is Basil Wright's 1936 *Night Mail.*

Paul Rotha, one of Grierson's associates, amplified and explained Grierson's thesis. He wrote, "... the film is photographed from real life and is, in fact, recorded 'reality' by the selection of images, brought about by an intimate understanding of their presence, the film becomes an interpretation, a special dramatization of reality, and not mere recorded description." Some years later, Rotha restated Grierson's definition of the documentary film as "The use of the film medium to interpret creatively and in social terms the life of the people as it exists in reality." Rotha's comments recognized the focus of documentary film as it was evolving in the mid- and late 1930s. Hardened through time, his definition precisely describes documentary film as we know it today. Significantly, Grierson first became interested in film as a medium for reaching the public, not as an art form.[5] Rotha's 1935 *Face of Britain* exemplifies this concept.

Psychological Distance. In 1933, Allardyce Nicoll discussed the concept of the "willing suspension of disbelief" in his essay "Film Reality: The Cinema and the Theatre." He contends that dramatic illusion is never an illusion of reality; rather, it is always <u>make-believe.</u> For aesthetic appreciation, the audience asserts a measure of psychological distance so they may "believe" (accept) what they see, even though they know it to be a dramatization.[6] Gerald. T. Rogers's 1984 film *My Father's Son* illustrates this point.

An apparent realism in characters, costumes, sets, objects, and forms creates an environment conducive to shortening the psychological distance. Conversely, in a surrealistic film, such as Jean Cocteau's 1930 *Sang d'un Poète* (*Blood of a Poet*), where the nightmare is realism played in fantasy sets, the audience increases the distance, never "believing," just being entertained or mystified from afar, having minimal or no empathy.[7] Significantly, some audiences who understand the nature of such a film will accept it for what it is and have an appreciation of it as an "artistic" achievement.

Appearing to be intrinsically at variance on first glance, these five principal concepts of reality in film, in fact, are mutually reinforcing. Each has a valid premise, and each makes a cogent point.

The Filmic Genres

The filmic forms of motion media range from theatrical extravaganza to lyric poem, from hard-hitting exposé to the fluff of a television comedy, from perceptual motor-skill training to the abstractions of Norman McLaren. From here onward in this chapter, I use the word "film" to encompass all types of motion-media.

I've categorized the entire gamut of film into five genres.

- Information
- Documentary
- Narrative
- Enrichment
- Experimental

I've defined each genre by the aura of reality its audience might perceive and plotted it on the grid in my Film Reality Scale. (Figure 12).

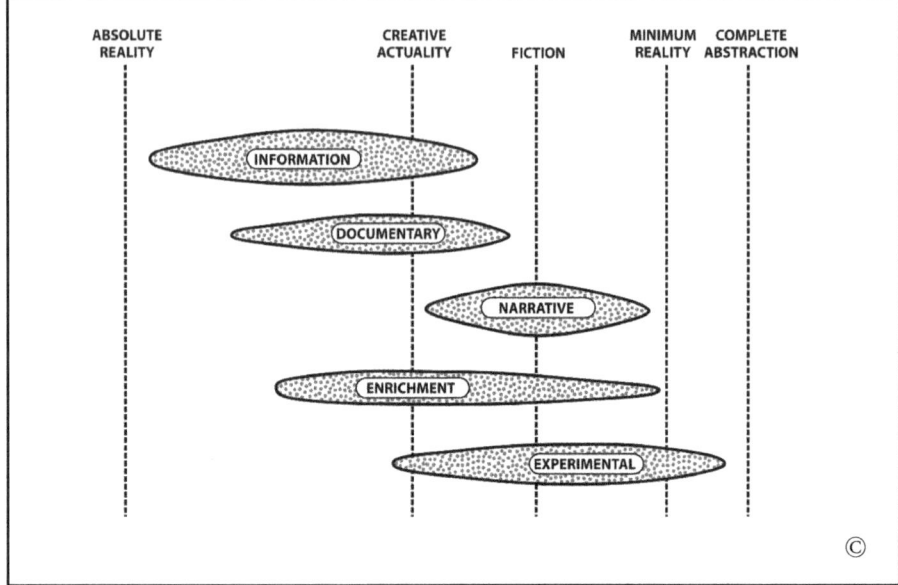

Figure 12. The Film Reality Scale.

I've devised The Film Reality Scale as a <u>nonlinear</u> continuum that depicts the gradual evolution of change of "film reality" and positioned the five cardinal measures on this continuum according to the inherent nature of the film genres. These points are:

- Absolute Reality
- Creative Actuality
- Fiction
- Minimum Reality
- Complete Abstraction

If I were to classify and plot every film ever made by a minute dot on the Film Reality Scale they would group, on the whole, into the irregular images representing the five genres.

Admittedly, there would be some dots scattered randomly throughout the diagram, because not all films fall into discrete genera.

There is considerable overlap among the film genres' diagrams because each type spans a broad spectrum on the Film Reality Scale. I've placed the termini for the film genres based on my subjective evaluation.

Absolute Reality. In the categorical meaning, "reality" can never be achieved in film. Increasingly, as we introduce cinematic techniques and the plastic elements of the medium, the film becomes more illusory and thus less "real." It's the audiences' psychological factors that engender their *willing suspension of disbelief.*

Creative Actuality. Located about midpoint is Grierson's concept of "creative treatment of actuality."[8] In creative actuality, the filmmaker has restaged (or staged) many of the scenes in the film. Much of the restaging reflects the filmmaker's bent toward emphasizing the goals of the film. As fanciful elements are introduced, the film drifts increasingly toward the fictional.

Fiction. As with absolute reality, there cannot be total fiction in film except perhaps in some bizarre fantasy film. Even in fiction (narrative) films, the actors themselves are real people speaking a recognizable language. My definition of "fiction" is that the film's major dramaturgical elements of plot, events, and characters are fictitious, while ancillary dramaturgical elements, such as props, costumes, locales, sets, and background, are actual, or nearly so, and the historical setting is reasonably accurate. As reality wanes in the major elements, the tenor of the film approaches the outlandish and, concurrently, as the ancillary elements become less real, the film drifts toward the phantasm of minimum reality.

Minimum Reality. At the point of minimum reality, communication, by its standard definition, becomes insignificant. The film becomes more an exercise in artistic achievement than a carrier of messages. It's a statement of self-expression. Surrealism with bizarre or incongruous symbols in unnatural juxtapositions prevails. As these characteristics intensify, the film evolves quickly to nonsense and complete abstraction.

Complete Abstraction. By definition, no film can be completely abstract, though some approach it closely. Near this point on the Film Reality Scale, the film has no meaning except perhaps to the filmmaker. The film is aesthetic gibberish, a fantasy of unrecognizable images linked by a totally random process of no significance.

Classifying Films by Genre

The elements I use in classifying film by genre are:

- Purpose of the film
- Primary audience
- Production design (the mise en scène)
- Scope of the show
- Filmic techniques
- Plausibility of the plot
- Credibility of the characterizations
- Actuality of the events
- Distribution scheme

With these elements as our guiding influence, we can classify a film in its genre. The five film genres are defined with the following general properties:

Information film is a mass-communication tool, one produced to communicate messages to mass audiences. The general goal is to influence audience members so that they will act or think to accomplish the goals established for the film—that is, what to do or think, how to do it, or why to do it. Usually, a client sponsors the film, and it is produced conventionally with a moderate to low budget, using standard filmic techniques. Occasionally, the film shows a sparkle of artistic merit. A nonpaying audience views information films on private and public channels—classrooms, training centers, or television. One classic example of the information film is Robert Flaherty's *Louisiana Story*. The Standard Oil Company of New Jersey sponsored this film and its goal was to inform the citizens of Arcadia that oil production in their area was beneficial to all.

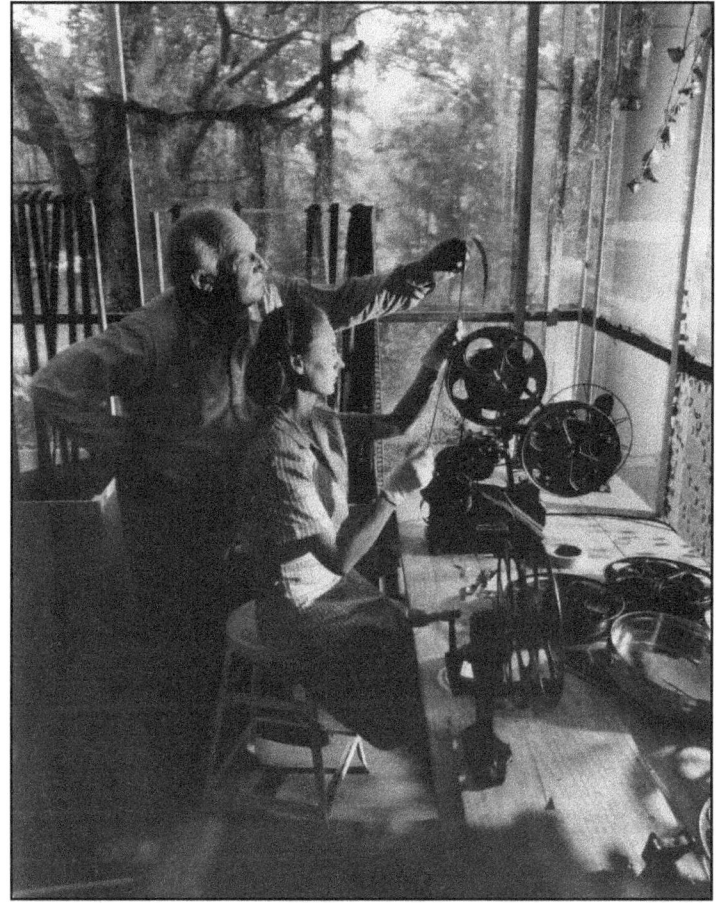

Figure 13. Robert Flaherty in Arcadia, Louisiana, working with editor Helen Van Dangen on his film *Louisiana Story.* Photograph courtesy of The Museum of Modern Art, Film Archives.

Documentary film is produced to enlighten the masses about a contemporary topic, a social idea, or an editorial position through the technique of *creative treatment of actuality*: a restaging or re-creating of events according to the filmmaker's interpretation of some facet of life and the world in which we live. That is, the film is a representation without introducing consequential fictional story interest. Production values are higher, on average, than those of the information film. Some such films are aesthetically accomplished. Generally, a nonpaying audience views a documentary film on the public channels of television. However, it is not uncommon for this film type to be seen in a classroom or in a theater (especially in Europe) where admission is charged, as it was in the beginning of the documentary-film movement. One sterling example is Lieutenant Commander John Ford, USNR's *December 7th*.

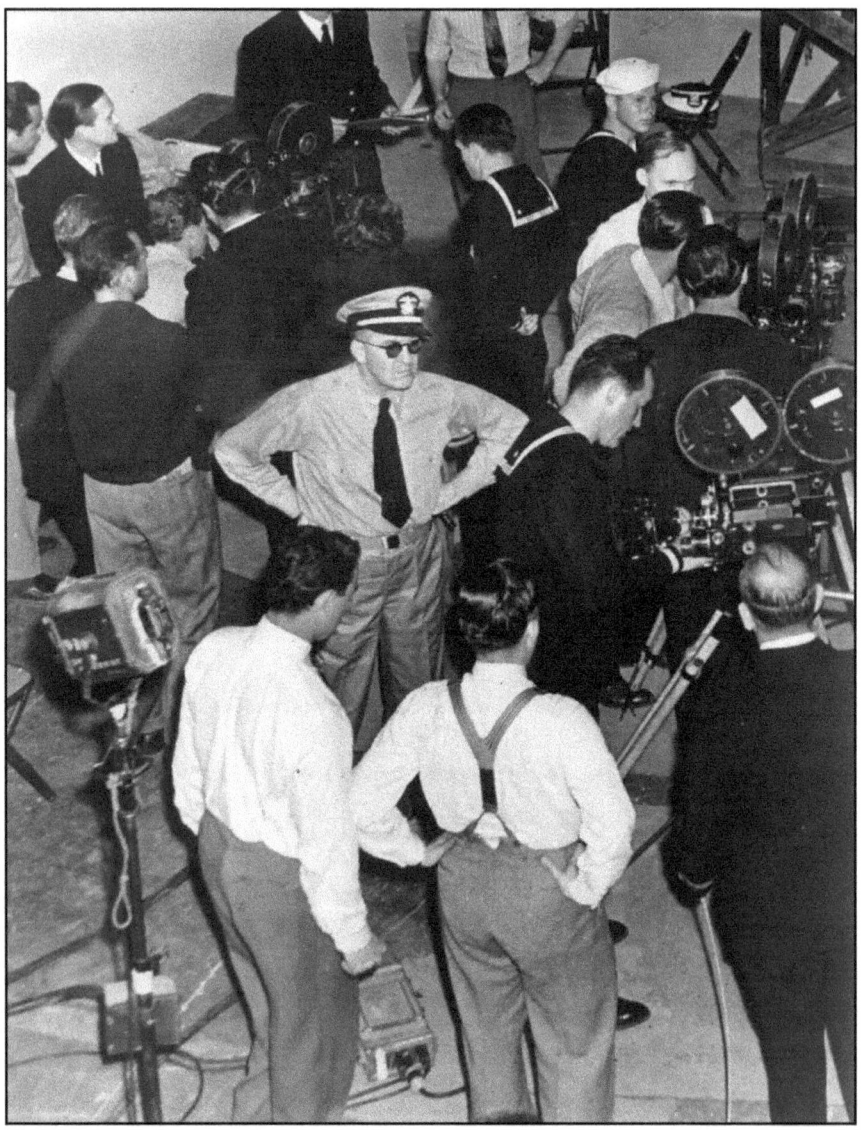

Figure 14. Lieutenant Commander John Ford, USNR, (the officer in khaki uniform and flat hat) directing a staged scene for his film *December 7th*. He is on the sound stage of the Naval Photographic Center, Anacostia, Washington, DC. Photograph courtesy of the Museum of Modern Art, Film Archives.

Narrative (entertainment) film is an entrepreneurial venture, produced to make a profit. Almost always, narrative film is a drama based on human emotions, experience, and conflicts, and many of the dramaturgical elements are fictional. Narrative film has very high production values compared with the other genres. We view narrative film in theaters, on commercial television, and in private settings. Narrative film provides enjoyment, pleasure, diversion, etc., to very large, heterogeneous audiences. Profit is realized by selling airtime, collecting admissions, rental fees, and royalties.

Enrichment film is difficult to classify because it's usually produced for aesthetic enjoyment and peer recognition. Generally, this film has no specific communication goal to achieve except in the broadest terms of a central theme. Production cost is recovered and profit may be realized from sales, rentals, or public screenings. Often, the theme is chimerical, production techniques are unorthodox, and the budget is limited. Usually the enrichment film is privately funded or financed by grants. Enrichment films are screened at film festivals and in museums, cine clubs, classrooms, and other private channels for cineasts and other selected audiences. The cognoscenti savor the show's artistry. At times, enrichment film may be seen on television and even more rarely in commercial theaters. One show that I've found to be especially cogent is Norman McLaren's *Pas de Deux*.

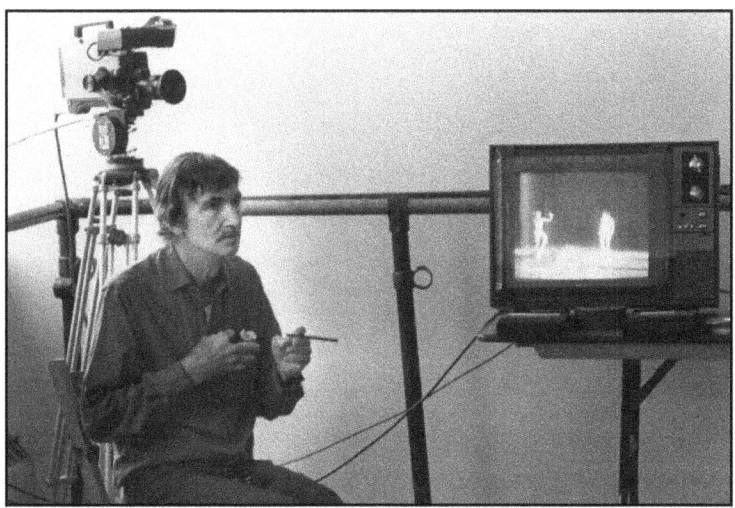

Figure 15. Norman McLaren works on a scene from his film *Pas de Deux*. Photograph Courtesy of the National Film Board of Canada.

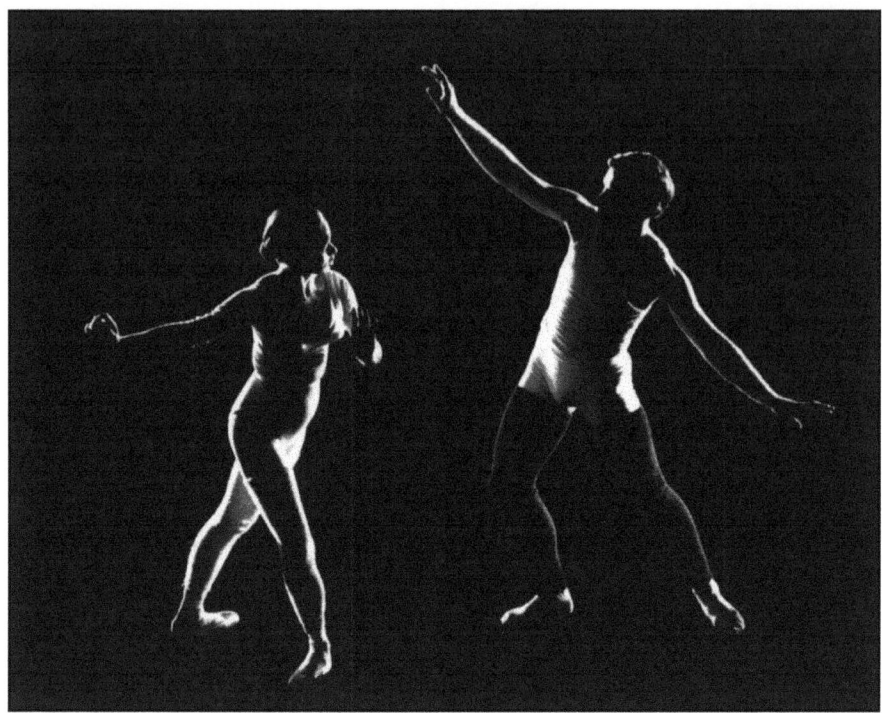

Figure 16. A frame from Norman McLaren's film *Pas de Deux*. Photograph courtesy of the National Film Board of Canada.

Experimental film includes several subtypes, such as abstract, avant-garde, fantasy, implausible, impressionist, surreal, and underground. In experimental film, anything goes, and it should. The imagination has free rein and is bounded only by the limits of the filmmaker's skill and the production budget. An experimental film is the filmmaker's statement of some inner compulsion—an interpretation of a dream, for example—expressed by a collage of imaginary images distorted in time and space, a grotesque illusion. Often, abstract visual patterns prevail in a fantasy of artistic experimentation not bound by any tradition and not produced to accepted norms. Nowadays, the experimental film is frequently privately funded, sometimes by grants. Often, experimental film is seen (and appreciated) by elitist film buffs and jaded aficionados in private or semiprivate screenings. Experimental film is screened also in art houses and noncommercial theaters for discriminating audiences. Film students write papers about the experimental films they screen and try to emulate, usually without much success. Particularly delightful is Ishu Patel's *Bead Game.*

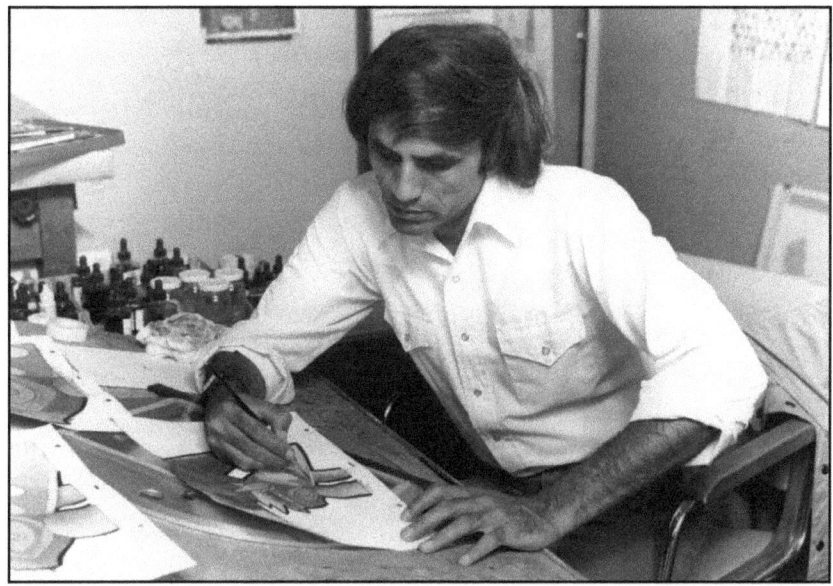

Figure 17. Ishu Patel designing an art scene for his film *Bead Game.* Photograph courtesy of the National Film Board of Canada.

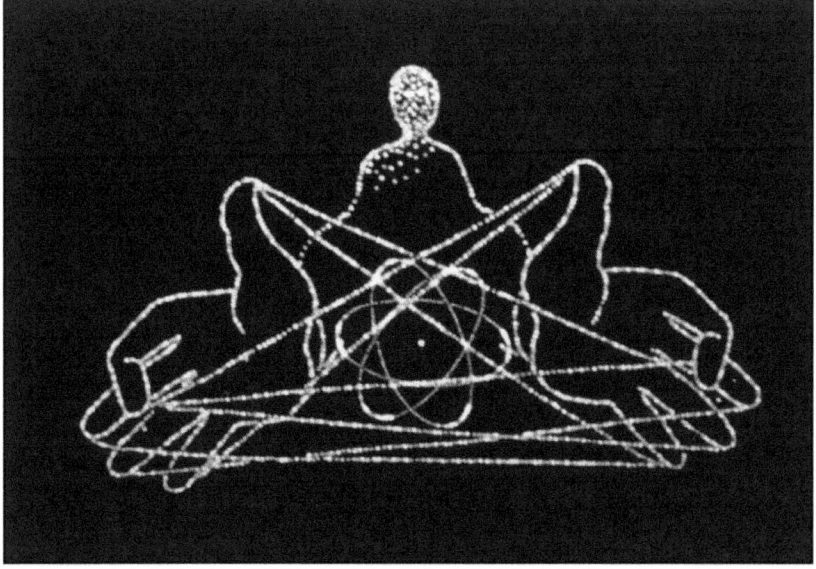

Figure 18. A frame from Patel's film *Bead Game.* Photograph courtesy of the National Film Board of Canada.

Of all the film types, it's the experimental film that performs the aesthetic research-and-development function of the medium—revolutionary ideas are tried, new cinema techniques are introduced, and old ideas and techniques are rediscovered, modified, and honed. Some music videos are examples.

Discussion

I admit that the classification of all films into these five types is subjective and reflects my predilections. I recognize as equally valid a host of other classifications and placements that would reflect another unique perspective. Paul Rotha, for example, classifies film into twelve categories, including, for example, Abstract, Cine-poem, Cine-surrealistic, Fantasy, Cartoon, and Documentary or Interest. Interestingly, Rotha does not recognize the information film. At best, he defines the film type he designates Cine Record "as a representation of modern fact, without the introduction of story interest . . . to be found in current newsreels." Another writer and filmmaker, J. Walter Klein, notes that there are "three general categories of nontheatrical films shown to the public: television commercials, sponsored film, [and] education films." He does not specifically mention these as belonging to the information-film type as I defined it.[9]

There are, of course, films that cannot be classified easily. These exceptions defy pragmatic classification because they span a large segment of the Film Reality Scale, having substantial elements of several types, or because they appear to be something they inherently are not. With minimal justification, these films could be placed in any one of several categories. For example, there probably is some entertainment value in all films, but not all films are produced for entertainment. In addition, almost all films have some informational value, though clearly all are not information films.

In the final analysis, it is the individual in the audience who makes the categorization, either consciously or, more likely, unconsciously. According to the social-categories theory, these categorizations are based on the individual's unique evaluation of what is real and what is fantasy—that is, what is plausible and what is sophistry.[10]

Let's explore this audience-perspective thesis with an example. A film with a deeply religious theme justifiably can be placed in any type; it depends on what member of the audience makes the decision. A devoutly religious person might classify the film as information and plot it near absolute reality. An agnostic might classify it as entertainment. An atheist might classify the film as experimental, approaching complete abstraction.

Conclusion

The effectiveness of a film is determined, in large measure, by the degree to which the audience becomes involved in experiencing vicarious identification with the characters, events, and places depicted in the film—its "reality."

Chapter 8

Documentary Film: A Learning Tool

If we are to excel as information motion-media designers and producers, we must have a deep and broad-based understanding of our profession—its theory, history, and philosophy. Such background knowledge, coupled with hands-on skills, constitutes professionalism. The hallmark of a professional is the insatiable desire to learn, to constantly expand and improve one's knowledge and skills.

One of the best ways to gain understanding and to increase our script-designing and producing skills is to learn from the masters, those filmmakers who designed and produced the classic documentary motion-media shows. If we're fortunate, after we've completed our education we'll serve an apprenticeship under the tutelage of a master—the hard and sure way to learn our profession. Under such tutelage, one of the most important skills we'll develop is the "critical eye," that uncompromising perspective that keenly discriminates between effective and ineffective motion-media, between communication craftsmanship and schlock.

Unfortunately, most of us don't have the advantage of serving an apprenticeship. Instead, we're plunged into the motion-media profession with potential but lacking a well-developed experience. We lack in-depth understanding, essential skills, and most important, the "critical eye" perspective. Thus, we don't discriminate as well as we ought. The quality of our shows suffers in a blur of personal and professional compromises.

I understand what's happening. We have little time for reflection as we respond to today's compelling demand to design and produce our

motion-media shows economically and quickly. We're seduced by new technologies that make our designing and producing tasks easier, faster, and
cheaper, but not necessarily better. "Better" is strictly a measure of audience
empathy and goal achievement.

Screening Classic Documentary Films

An excellent way to develop the critical eye is to screen the documentary
films of the master filmmakers. Screening is not enough, however; we must
analyze and critique them with our developing critical eye. We are directed
toward understanding how the masters accomplished the communication
task at hand through relevant filmic design—the grammar and syntax of the
medium. Questions to ask are:

- Does the film have a clear vision and tone that communicate a strong
  central theme?
- Does the film generate empathy in the target audience?
- What filmic-design techniques are used and why are they effective?
  Ineffective?
- Is most of the information encoded in the kinetic visuals?
- Is there coherence between the narration/dialogue and the kinetic
  visuals?
- Is the pace appropriate?
- Does cinematography/videography facilitate communication?
- Are music and sound effects used appropriately to enhance the
  scenes?
- Is the plasticity of the medium (manipulation of time and space)
  used optimally to maintain orientation and to facilitate information
  flow?
- Are spatial relationships and art elements used to accentuate import-
  ant points?

In Appendix One, "Evaluating Information Motion-Media," I've included
my "Motion Media Evaluation" scoring form, a few comments, and instructions. I recommend that you use this form to evaluate motion-media

shows for their communication effectiveness and to develop "the critical eye." (Page 190.)

During your critique, it's best not to concentrate on negatives. Too often, such activity degenerates into time-wasting, nitpicking exercises of no real import, and the essential goal of understanding is lost. No filmmaker ever made a "perfect" film, and no filmmaker is totally satisfied with his or her film.

Documentary Film Background

If we are to appreciate documentary films to full measure and make relevant analyses, it's important that we have some background on the documentary film. This discussion will not be an in-depth review of the history and significance of this genre. Rather, what follows is a brief discussion of what a documentary film is.

Documentary film pioneers such as John Grierson, Paul Rotha, Eric Barnouw, Lewis Jacobs, Roger Manvell, and Pare Lorentz have written in-depth commentaries on the topic. Modern-day authors who expound on documentary film include Rachel Low, Elizabeth Sussex, Eva Orbong, and Forsyth Hardy. (Please see the bibliography for a complete listing.)

Most of us in the information motion-media profession consider John Grierson, Commander of the British Empire (1898-1972), to be the founder of the documentary-film movement. In 1927, under the aegis of Sir Stephen Tallents (1884-1958), Secretary of England's Empire Marketing Board, Grierson founded the Empire Marketing Board (EMB) Film Unit.[1] Under Grierson's tutelage, the group developed and refined the documentary film to a fine art, producing a host of classic documentaries over the years. I've included several of these documentaries in the list in Appendix Two, "101 Classic Documentary Films." (Page 193.)

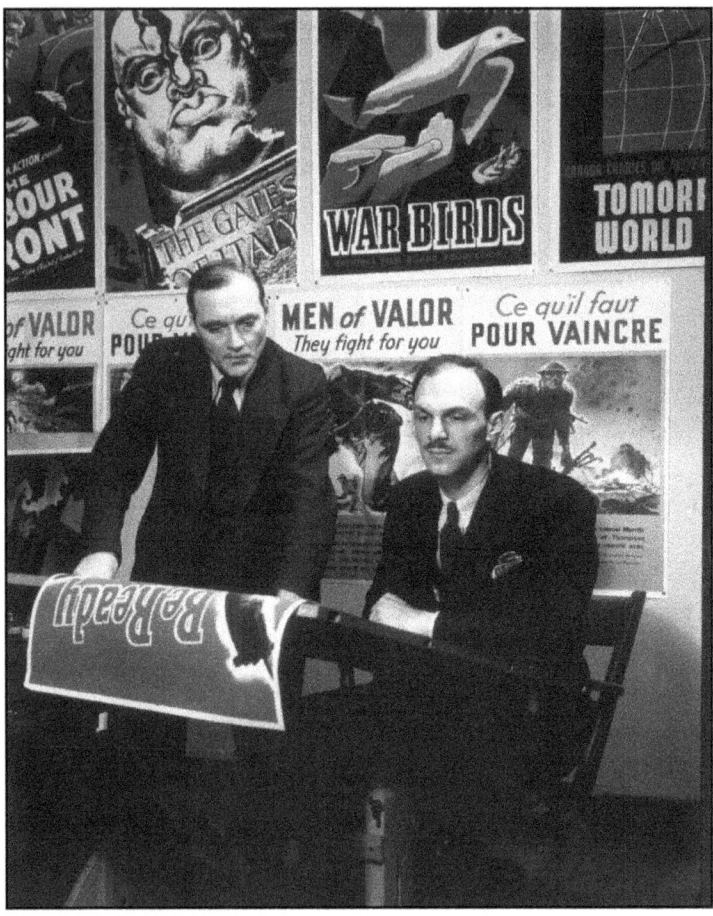

Figure 19. On the left is John Grierson, Commander of the British Empire (1898 – 1972), with an unknown assistant, working in the Board's studio in Montreal. Photograph courtesy of the National Film Board of Canada.

It was Grierson who first applied the term "documentary" to a film. In an article in the *New York Sun* in 1926, Grierson reviewed Robert Flaherty's film *Moana: A Romance of the Golden Age*: "Of course *Moana* being a visual account of events in the daily life of Polynesian youth, has <u>documentary value</u>." (My emphasis.) Grierson defined the documentary film as "the creative treatment of actuality," which acknowledges that in documentary filmmaking on "the living article, there is also opportunity to perform creative work." Roger Manvell amplified this concept as the filmmaker's interpretation of a factual subject as the filmmaker sees it.[2]

Complementing Grierson's definition, Paul Rotha, a sometime associate of Grierson in the General Post Office Film Unit (the successor to the Empire Marketing Board Film Unit), postulated that the origin of documentary film lies in Robert Flaherty's *Nanook of the North*.[3] Produced in 1921 for Revillon Frères of New York, the film tells the human story of Eskimos in the far northern climes of Hudson Bay. Much of the cinematography was staged for dramatic impact, or "reconstructed," to introduce creative art into the recording of Inuit life on film.

Grierson developed the documentary film's first principles:

- "Documentary would photograph the living scene and the living story.
- We believe that the original (or native) actor and the original scene are better guides to a screen interpretation of the modern world.
- We believe that the material and the stories thus taken from the raw can be finer (more real in the philosophic sense) than the acted article."[4]

The documentary tag stuck. And in the intervening years, "documentary" was applied to a host of realistic-type films—indeed, to almost anything that was not theatrical or narrative.

Significantly, Grierson insisted that the primary aims of documentary film are national education and public information, and that the documentary should strive for sociological rather than aesthetic goals. He insisted that within the documentary film, "there must be power of poetry or of prophecy. Failing either or both . . . there must be at least the sociological sense implicit in poetry and prophecy."[5] It is important that Grierson did not pooh-pooh aesthetics. During the 1939-1945 war, Grierson's former documentary film group (now the Crown Film Unit) fine-tuned the documentary film into a propaganda medium of a high order with such classic films as *Target for Tonight* and *Listen to Britain*.

Through the years, other documentary filmmakers have augmented Grierson's definition and philosophy with their own perspectives. Paul Rotha said of the "Cine-Record" documentary film, "The representation of past fact, without the introduction of fictional story-interest, is an attempt to put on record the actual happenings of some past event."[6]

Ivor Montagu, film director, critic, historian, and pamphleteer, tend-
ed to de-emphasize human interaction in documentary film. His is a more
pragmatic approach in which the nonhuman objects in the scene are of primary importance. He said, "Documentary deals with nonhuman objects and
processes, or processes in which, if human beings are included, this is only
in relation to their offices and functions and not in respect to their qualities
and interrelationships as individuals." What impressed Montagu was "the
cinema's capacity to present a plain scientific record [of man]. . . . No other
medium can portray real man in motion in his real surroundings."[7]

Perhaps it was Pare Lorentz whose simple yet elegant definition is the
most profound. He described the documentary film as "factual film which is
dramatic."[8] Such a definition is reflected in his Great Depression-era films,
such as *The Plow That Broke the Plains* and *The River,* which set the documentary standard in the United States of America.

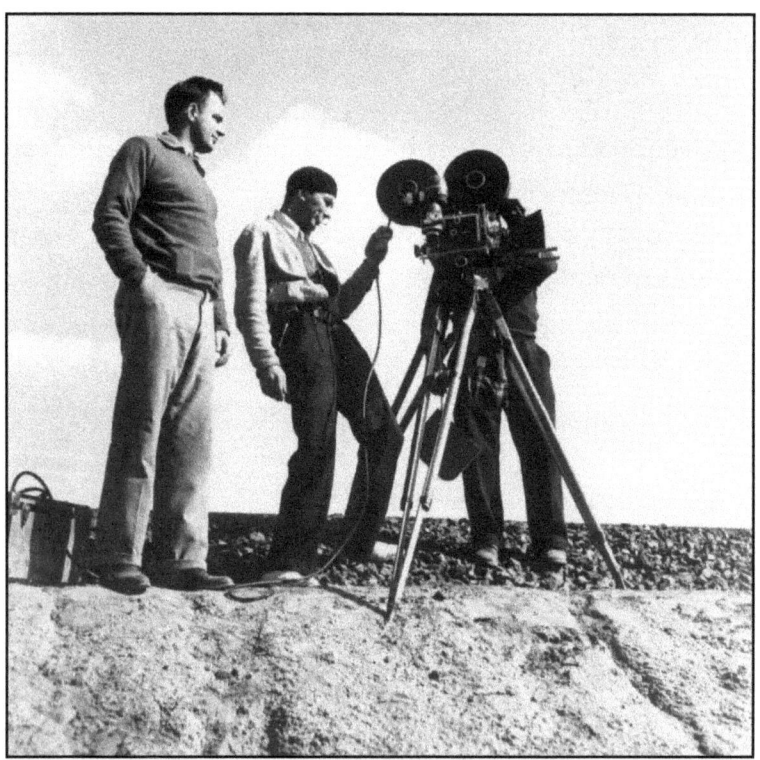

Figure 20. Pare Lorenz directs his cinematographer, Paul Ivano, in a scene from his
film *The Plow that Broke the Plains.* Location is Bakersfield, California. Photograph
courtesy of The Museum of Modern Art, Film Archives.

Over the years, documentary film has been a key element in influencing mass audiences, particularly about social themes. Today, the genre has been refined to a hard-hitting communication art form that, on occasion, can take a position of extreme advocacy or exposé that is blatantly nonobjective. Two-time Academy Award-winner Robert Richter produced documentary films in advocacy style. One example is his 1980 film *Pesticides and Pills: For Export Only*. He charges several multinational corporations with knowingly selling dangerous pesticides and pills in Third World countries where consumers lack the sophistication to know what they are buying or how to cope with it if they do know. Other current-day documentary filmmakers take a less extreme position, following more traditional lines—Ken Burns' 1990 *The Civil War*, for example. However, social themes are common to all in varying degrees of intensity and importance.

Documentary Film List

Without formal education or extensive background, it's difficult to know who the documentary-film masters are or to know their work. To this end, I've compiled a list of 101 classic documentary films in Appendix Two. (Page 193.) They're listed in alphabetical order. I've listed films with non-English names by their original titles.

Some readers may note that I've omitted important films and filmmakers. Others may question my choices, preferring a different selection. Admittedly, the list I've compiled is incomplete; my subjective preference and space limitations dictate the list's content. Yet the list is a representative sample of classic work that spans a broad spectrum of styles, goals, and milieus. The films are cosmopolitan, reflect varied political ideologies, and range from contemporary to *vaunt-courier*.

Not all of the films on the list are, strictly speaking, documentary. Some are information films. A few others are enrichment films. A couple are experimental. The common thread in all the films is that they are not narrative films. In the broad context, therefore, we can consider all of them a part of the documentary genre. Such a conclusion is especially valid when considering Grierson's support for inclusion of a host of film types within the documentary genre. He said, "Think of all the different categories of documentary productions, e.g., public reporting, scientific films, technical films, instructional films, etc."[9]

Many of the early documentaries may well appear crude by today's standards. Contributing to this perception are technical inadequacies, unsophisticated filmic techniques, and film's natural deterioration over time. Don't be fooled! Look through these superficialities to see the real worth of these films—their poignant power. Focus on flaws, and you'll miss gaining a basic understanding of the origins of the documentary genre and an appreciation of the pioneers who produced these films. Discover the filmic-design techniques of the masters. Hone your perspective on this powerful communication medium, nurture that developing critical eye—our key to professional success.

To find copies of these shows, try the Internet. Search university and public libraries, museums, and film-archival institutions. One excellent source for film and videos held by university libraries is *Educational Film/ Video Locator* by the Consortium of University Film Centres, published by the R. R. Bowker Company of New York.

Note. On 19 January 2019, I spotted this book for sale on Amazon.

Shelton's Pronunciamento
Review documentary film with a critical eye.

Chapter 9

Introduction to Multimedia

Multimedia is the vanguard of the media revolution: it's powerful, *nonlinear*, and most important, it's <u>interactive</u>. Multimedia is a computer-based medium. And it's the computer that begot the media revolution.

Definition

Technology is developing so rapidly that we can't pin down a definitive description of multimedia. By the time you read this, I'm confident that some (or most) of what I've said here will be obsolete, and some new multimedia technology will have evolved. That's one reason I don't discuss tools. Accordingly, this chapter necessarily treats this complex subject superficially. Nonetheless, here's my tentative definition:

Multimedia is an umbrella term that describes an information medium that is often designed and controlled on a computer. The software integrates various external visual and audio inputs and computer-generated graphics into an integrated sight-and-sound, nonlinear, interactive show.

The viewer controls the rate of information flow and its sequencing. Accordingly, such shows mimic the user's associative thought and cognitive learning processes by requiring viewer selectivity and response. Real-time feedback is an essential advantage of multimedia.

In figure 21, I've drawn a generic composite of the various elements that comprise multimedia. The figure is not intended to be all-inclusive or detailed. Rather, it's a broad overview of the media inputs, computer functions, and program outputs.

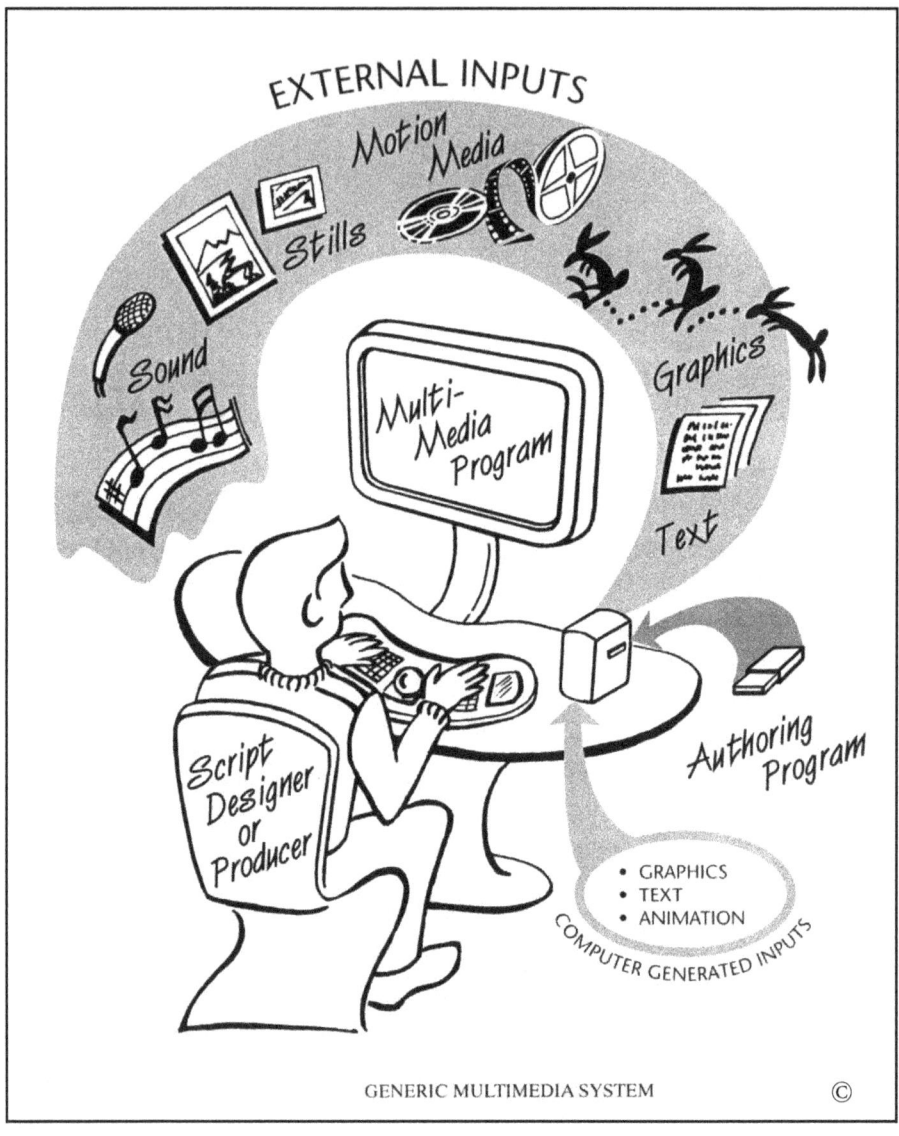

Figure 21. A generic multimedia system.

Interactivity is the key to the communication power of multimedia. It allows for one-on-one interaction between the viewer and the multimedia presentation—the instructor. Students assimilate information at their own pace. They can browse, review, scan, freeze-frame, fast-forward, or reverse. A number of studies prove that multimedia significantly improves learning in the short- and long-term for all audiences. Such viewer control of information flow is especially useful in teaching subjects whose nature is linear, that is, sequential; for example, teaching perceptual motor-skills or language mastery.

Background

I am not sure when multimedia was born, the term or the process. I suspect, however, that it began in the late 1970s with the rise of computer-aided instruction and multi-image slide shows. Then, the technology supporting multi-image shows consisted of an array of 35mm slide projectors, stereophonic reel-to-reel or cassette tape-recorders/players, and maybe some 16mm motion-picture film. All elements of the show were controlled by various electronic "black boxes." I've seen shows with twenty-five projectors and have heard of shows with more than forty. As you might imagine, multi-image shows were plagued with technical snags—stuck slides, burned-out projector bulbs, loss of sound synchronization, etc. Multi-image had a meteoric rise and concurrent decline because the medium could not overcome the technical complexities and distribution logistics.

Soon the "black box" became the personal computer. Advancing digital-compression technology enabled video and audio information to be stored in a minute fraction of the space once required. A new generation of external and internal plug-in devices further expanded computing capabilities. Now, the personal computer could generate complex 3-D graphics, animation, and all manner of visual effects. Concurrently, multimedia-authoring programs became more generic and powerful. Such programs integrated a host of visual and aural inputs with computer-generated information into one medium. For the first time, a broad spectrum of our audience had access to an interactive medium that was practical, cost-effective, and readily available.[2]

We're now at the point where technology has made it possible for just about anyone who is computer literate to produce a multimedia show. The question is, does the "anyone" have the in-depth knowledge, superior filmic-design skills, artistic ability, and panache to produce an effective show?

Advantages of Multimedia

Viewers' self-paced activity is the basic strength of multimedia—increased learning and learning more quickly. A 1987 study by the Gillen Interactive Group indicated that students using interactive programs learn and retain 25 percent more of the information presented and learn it 50 percent faster than those who use traditional learning methods. A series of six studies by Nathan Kolowski, conducted from 1990 to 1992, showed that students using multimedia have a 55-percent learning gain over traditional classroom teaching. Students learn the material 60 percent faster and their long-term (thirty-day) retention ranges from 25 to 50 percent higher.[3]

Let's review the advantages of multimedia compared to linear media.

- Multimedia combines the power of self-study with computer management in one medium.
- Viewer has near-immediate access through branching networks to any part of the show that allows for individualized, self-paced instruction.
- Viewers control the rate of development through such options as forward, reverse, freeze-frame, and stop motion.
- Complex concepts are explained via animation, motion video, still photographs, and drawings.
- Narration and dialogue can be heard in any of several languages.
- Viewer performance can be collected by automatic record-keeping. Such data is used to determine user accomplishment, program strengths and weaknesses, frequency of repeat viewing, and times of use, for example.
- Viewers' questions are answered in near-real time through internal communication networks.
- Hazardous-training situations can be simulated: for example, bomb-disposal training or firefighting.
- Expenses for training-related travel are reduced. Send the multimedia show to the student/viewer.
- Training (viewing) can be scheduled to meet the needs of the viewers.

- In many multimedia training and education scenarios, an on-site instructor is not needed.

Virtual Reality

Virtual reality (VR) may be the next-generation software for training, entertaining, and whatever. VR is that computer-generated technology which creates a multisensory, computer-modeled environment of three-dimensional, holographic images with realistic multiphonic sound and lifelike tactical sensations. It allows for near-total sensory immersion in a realistic, 360-degree world of stereo-optic vision, sound, and touch. To view a VR show, the person uses special electronic equipment: a helmet with a screen inside and gloves fitted with sensors. When we are absorbed in virtual reality, it's easy for us to lose conscious perception of the real world.[6]

Current applications for virtual reality include sports training, surgery, medical diagnosis, flight simulation, war gaming, engineering design, 3-D layout for building design and furnishing, and perceptual motor-skill training of all sorts.

Shelton's Pronunciamento

Interactivity is the future in motion-media communication.

Chapter 10

Multimedia Flowchart

Our client has tasked us with devising a motion-media show to introduce junior-high students to four Florentine Renaissance artists and samples of their work. Expected student testing goals are 100 percent before quit, 85 percent in thirty days, and 65 percent in sixty days. Goal details are noted in scene 2 of the script, on the page shown below.

We've concluded that our client's communication objective is best fulfilled by a multimedia show. Accordingly, we'll need to create a flowchart. A flowchart is somewhat like a script or storyboard in that it's the map for our show that details what will be shown and said, the branching network, and decision points. Page one of a very simple multimedia flowchart is shown in Figure 22.

Let's explore this flowchart to follow the network branching, and learn what the symbols mean. Note the linear progression from the show's title to the map. The arrowhead lines indicate the direction of information flow and the network-branching options available to the student. Rectangles indicate a scene. Here, the scenes are text and graphics. The numbers inside the circles attached to the rectangles indicate the number of seconds the scene is on the screen. Significantly, the student, in this phase, does have the option to freeze the scene or to extend it. Note that when the show opens, the student has no decision options until after the map scene ends. The diamond with the notation "D1" is the first decision point. The student can choose to continue into the show or go back to any one of the opening graphics. The inverted trapezoid with the "2" is the connecting link to page two.

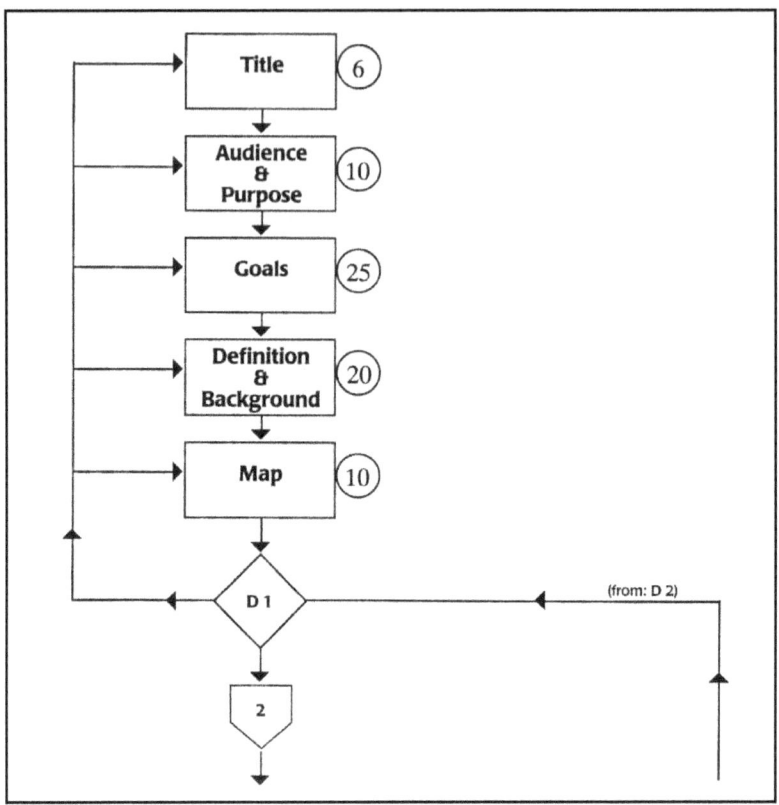

Figure 22. Page one of a multimedia flowchart.

To keep the flowchart easy to follow, I've left off the all-important *Escape* icon—it ought to be prominently shown on each page. Escape lets the student return to "Home" or to another familiar page. It's easy to get lost even in a moderately complex show. I've done it several times. Several years ago, I was at the Defense Language School in Monterey, California, reviewing a multimedia show designed to teach the German language. I was in the "Beginning" sector. Learned a lot quickly. However, if not for that "Escape" function, my lesson would have been a disaster.

Following is the script of this opening sequence.

Multimedia Script

Multimedia Script
A Sample of Florentine Renaissance Art

Visuals	Audio
(POP ON title) *A Sample of Florentine Renaissance Art*	(Music in) (Appropriate 15th century chamber music)

(Audience and Purpose)

	(Music down) (Narrator)
1. Montage of statues by famous Florentine sculptures. Use CLOSE Ups to emphasize the texture and form of the art. Let the scene play for several beats.	This multimedia show is an introduction to Florentine Renaissance art in the fifteenth and sixteenth centuries. We'll focus on the art of the masters sculptors.

(Goals)

	(Continue music softly) (Narrator)

(POP ON titles sequentially)
2. Titles over a neutral background.
  • Define Renaissance art
  • Identify the centuries in which Florentine Renaissance art flourished
  • Identify four artists
  • Name two works by each of the four artists.
	Shown are the minimum goals you are expected to achieve upon completion of this multimedia show.

(Definition)

	(Continue music softly) (Narrator)
3. Build a montage of examples of classic Roman and Greek sculpture. . Let the scene play.	Renaissance art is a derivative of the form and aesthetics of classic Roman and Greek sculpture.

(Background)

	(Music continues softly) (Narrator)
4. Build a montage of modern-day Florence city scenes: *Ponte di Vecchio, Piazza della Signoris, Il Duomo*, etc.	Renaissance art began in Florence in the early fifteenth century.

	(Music continues) (Narrator)
5. Montage of Florence's famous museums: *Galleria degli Uffizz, Galleria dell' Academia*. Feature Michelango's "David" and the "Slaves," and *Galleria dell Bargello;* for examples.	It flourished under the patronage of Florentine nobles who were interested in pre-Christian cultures.

(Location: needed for orientation)

	(Music is 20th century, Luigi Denza compositions, eg.) (The narrator may be quiet. Or you the reader may develop the narration for this scene, if any. Necessary?
6. POP ON map of Europe. Italy is highlighted. CUT To a MEDIUM CLOSE UP of Italy. POP ON a circle surrounding Florence on the map and POP ON the title "Florence."	
	(Segue accordion music as from an Italian cabaret)
7. Montage of modern-day street scenes of Florence.	(Narration? If so, what?)

(Script continues)

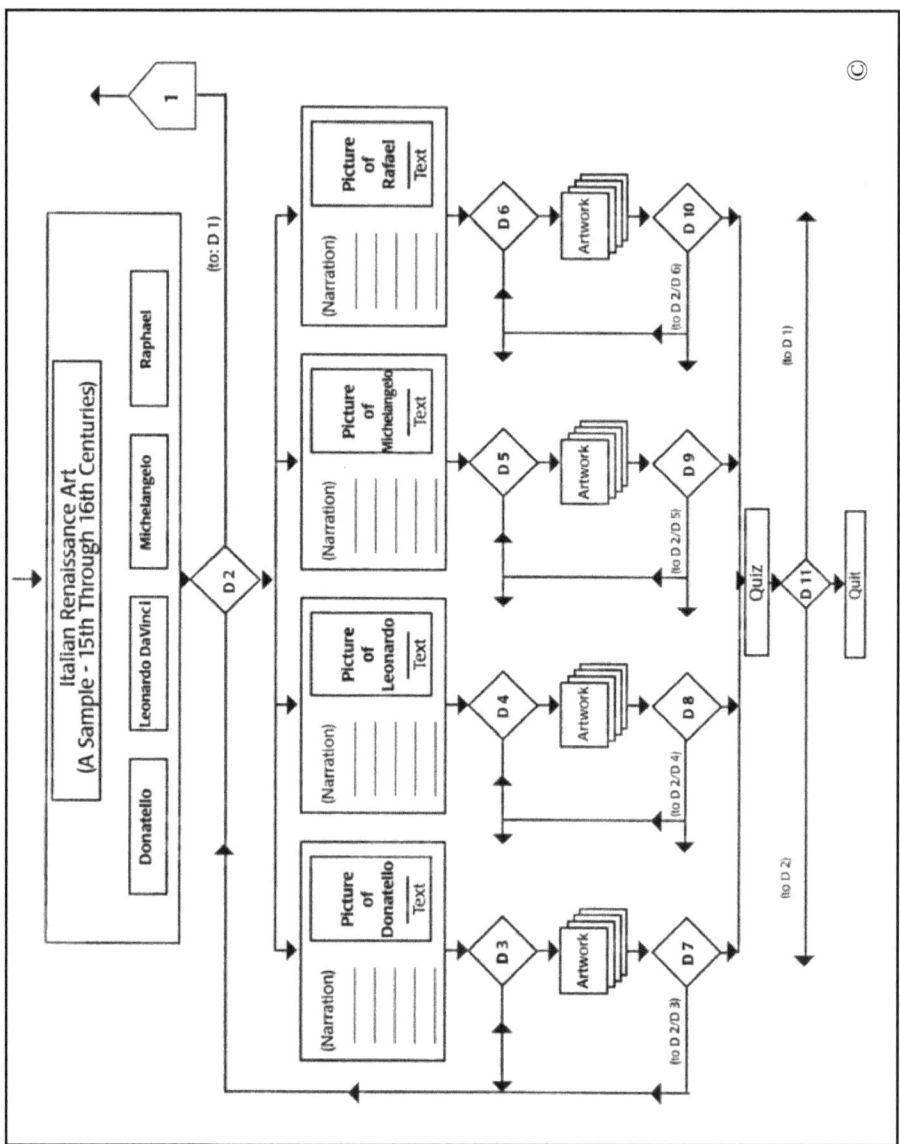

Figure 23. Page two of the multimedia script flow chart.

Note the many decision points and the complex branching networks in this 'simple' flowchart. I recommend that you trace it through to see all the options available to the student. Note the overlapping artwork windows. There, we could provide a requirement that students make side-by-side comparisons of the artwork of the different artists. Insert a quiz: pass or fail leads to another set of networks. Most all options for student involvement are available. Nonetheless, my golden rule is, make your flowcharts as simple and direct as possible—just intricate enough to optimize student learning.

Near the bottom of the flowchart is "Quiz." If students complete the quiz successfully, they might have the option to quit or to go back into the show to review a section. Should students fail, we can route them back into the show at any point of our choosing. We might base our decision on the specific questions that the student answered incorrectly, to force a review of that section of the show. And, we could insert quizzes at any part of the show. At the bottom of the flowchart is "Quit," which takes the student completely out of the show.

Thoughts on Copyright

When designing a multimedia show, it's tempting to copy kinetic art, still pictures, animation, and music of all types from the Internet.

Don't!

Almost all of these images and music are copyrighted. If you use them without permission, you're stealing another person's property. You're in violation of federal copyright law, and you may be prosecuted in federal court. Also, you're committing a tort. It's a sure bet that the copyright holder will sue you and your organization. It's not worth it. On page 165, Chapter 20, "The Sound Track," I highlight a few aspects of music copyrights.

Copyright law is beyond the scope of this book and of my expertise. Suffice it to say, you should consult expert legal advice to ensure that you have legal permission to use all the images and sound you include in your shows. *No exceptions!*

Shelton's Pronunciamento

Using images, text, or music without permission is theft.

Part III: Communication Analysis

Chapter 11

Stating Your Objectives, or, What's This Show About?

When the room lights come on after a motion-media screening, the audience should think or act in a way that accomplishes the show's goal. If so, then our show was successful.

Essential to the production of a successful information motion-media show is an in-depth communication analysis. Its three key components are

- Establish the show's goal
- Define the audience
- Detail the objectives

What's the Real Goal?
We can't take at face value what our client says is the goal. Sometimes, I've found that my client's idea of the goal, for whatever reason, is faulty. After conducting an in-depth analysis of the problem the show is supposed to solve/ameliorate, we might conclude that the actual goal for the show ought to be something other than the client's notion. Accordingly, before we begin topic research and script design, we need to understand precisely the actual problem, define the target audience, resolve the goal to be achieved, and de-termine the appropriate media in which to encode the message.

For instance, some years ago a client of mine, the owner of a small prototype machine shop, asked me to produce a show that would motivate his machinists to work more safely—a practical goal. Over the years, several machinists had had on-the-job injuries. Easy enough, right? Not so! After I visited the shop and talked with some of the workers, I concluded that the problem was not with the workers but with the industrial layout and inadequate lighting. My client should have consulted an industrial or safety engineer first. This problem didn't need a motion-media show fix.

Once we understand accurately the proposed motion-media's task, we must state the show's goal in terms that are <u>fitting</u>, <u>realistic</u>, and <u>worthwhile</u>.

- Fitting: The goal is appropriate for the problem at hand and the medium chosen.
- Realistic: The show has a reasonable chance of achieving the goal
- Worthwhile: The goal serves a useful purpose and is worth the funds, time, and resources needed to design and produce the show, and it's worth the audience's time, energy, and concern to view it.

An integral element of a show's goal is the audience's rate of achievement in the near term and long term. Ideally, we'd want a statistical measure of audience performance obtained through written and performance testing—the only sure way to know how our show fared. Unfortunately, we don't get much feedback on audience achievement in linear shows. Most multimedia shows score audience achievement in real-time testing and feedback. If our show fails, we can count on our client letting us know. However, success usually begets silence from our clients.

Goal and Audience

A statement of a show's communication goal is not complete until we couple it with the sociological analysis of the target audience. Who will constitute our audience and why do they need to view this proposed show? (In Chapter 12, starting on page 109, we discuss audience analysis in depth.)

To illustrate, I'll describe a motion-media communication problem I worked on some years ago. There were several horrendous accidents aboard the Navy's aircraft carriers operating in Southeast Asia because of the unsafe work habits of sailors handling liquid oxygen (LOX) on the flight deck. My Navy client wanted a show that would "train" the LOX handlers to work safely. LOX handlers, in those days, typically were eighteen- to twenty-year-old males.

After conducting extensive research, we found that the problem was not lack of training. Rather, it was one of negligence due to fatigue and the ennui resulting from long and exceptionally hazardous hours working under the most demanding conditions. All LOX handlers had attended a special school with rigorous standards and had received follow-up on-the-job training on the "how" of their jobs. Some of the LOX handlers had become inured to the hazards in handling this highly dangerous, volatile, and erratic cryogenic liquid. They "knew their jobs backward" and with staggering overconfidence had concluded that nothing would ever happen to them. Familiarity breeds contempt, as it were.

My task was to *motivate* LOX handlers to work with LOX precisely as they had been taught and as detailed in the LOX operator's manual. The goal was to reduce hazardous activities to **zero**. The solution to this problem was the highly effective, six-minute, award-winning motivational film, *The Man From LOX*. This show met its objectives to a fare-thee-well. LOX accidents dropped to **zero** and remained at zero for years afterward. Within a few years, all the NATO countries and most of the Latin-American countries adopted this show.

The Goal Statement

It is not sufficient to make general goal statements such as: to inform the public of pending legislation, to improve morale among employees, to increase sales, to teach sailors how to handle LOX, or to increase donations to Acme charity. The purpose must be specific and pose the audience's action item in concrete terms. Inherent in this statement of purpose is the key element that describes the target audience. With these factors established precisely, we've completed the first step in designing a successful show.

Let's review goal-statement development from another perspective: motion-media competitions and festivals. It's here that we see what sorts of shows our peers are producing and what success (or not) they're having.

I managed a major international motion-media competition for twelve years, served on a host of local and blue-ribbon juries for several other national and international festivals and competitions, and I screened as many information shows as I could. In almost all festivals and competitions, the scoring is based on the judges' subjective evaluation of the show. In some competitions, the scoring weight for achievement of objectives (and related criteria) can be as high as 50 percent of the total score.

Other factors considered are "creativity" (not defined), and the various aspects of technical competence that contribute to goal achievement. (See Appendix One, "Evaluating Information Motion-Media." Page 187.)

Time and again, I've seen "well-produced" shows score poorly because the statement of objectives is inadequate—although not "inadequate" in terms of brevity (if it were only so!). Rather, an inadequate statement of objectives fails to tell precisely what are the show's goal and target audience. Such an inadequate statement of objectives guarantees that the judges (and most likely the target audience) won't know what a show is all about—a fatal flaw. An inadequate goal statement almost always reflects incomplete communication analysis. For instance,

- Lack of clear understanding of the problem the show is designed to solve
- Failure to articulate the target audience's achievement
- Inappropriate filmic design

Consequently, the show's communication effectiveness suffers from lack of relevance and focus. This kind of show tends to wander, is often badly overwritten (to cover all contingencies), and usually doesn't do much for anyone in the audience. The technical razzle-dazzle may be great, but on reflection the judges (and audience) wonder what the show was all about.

Let's get personal. Is it possible that our "great" shows haven't fared as well as they ought to in competitions because of an inadequate statement of objectives?

Let's earn an award at the next competition. It's fulfilling to garner peer recognition, publicity, and kudos. Such recognition gives us a sense of self-worth and accomplishment—makes our jobs worthwhile. Management and clients take pride when one of their own earns recognition. Festival and competition awards are evidence of your (and your group's) talent and importance—a kind of built-in job security, as it were.

Unfortunately, I see far too many shows entered in competitions and conferences with insufficient statements of objectives—perhaps as many as 85 percent. The insufficient statement of objectives comes couched in many guises: some obviously fatuous, and some deceptively void of any real goals.

I've collected the following potpourri over the years. Most are quoted verbatim, others I've modified slightly to illustrate the point.

- *Ambiguous:* "We hope this film does everything for our client that she expects of it. So far, she has been very pleased because the audiences seem to like it," etc., etc., etc.
  - ° What's "everything"? And "like" *don't* count!
- *Irrelevant:* "I'm so thrilled to enter my 'wonderful' video in your swell competition. Watch for . . . and see how we accomplished . . . and we tried so hard to . . . and our budget was so small . . . and I just know you'll love my wonderful video, and. . . . "
  - ° Quoted verbatim, unfortunately.
- *Worthless:* "I'm concerned about how you fellows judge my shows. Here's my new film to let you know just what a great filmmaker I am."
  - ° Folks, I kid you not. This is an exact quote without the invectives.
- *Rambling:* "We had this kind of problem at work, and our public relations director thought we in the Communications Division could help. She called us . . and we produced this show. . . ."
  - ° That's wonderful. Hope you had a fine time on the company's money. By the by, what problem?
- *Ad nauseam:* One statement of objectives rambled for three typed, single-spaced pages!
  - ° That's tawdry and incompetent. The other judges and I were not motivated to read this tome. We passed. The show '"bombed."
- *Synoptic:* "This show is about how Jane and Jim cope with and overcome the problems of poverty in their newlywed lives."

- ○ This sort of statement of objectives is nothing more than a capsule summary of the show. The judges don't need a synopsis; they need to know what's supposed to happen when the lights come on.
- _Incomplete:_ "This film explains how a high-powered laser will be used to produce an unlimited supply of affordable energy."
Here's another: "To provide an overview of the integrated-services digital network."
  - ○ Generally, these kinds of statements of objectives are okay as far as they go, but they don't go far enough. They don't pinpoint the target audience and what the producer expects from the audience.
- _Litany:_ "In this video, we tell the audience about the details of processes A through Z, how to integrate the processes, why the processes need to be integrated, what the results of such an integration are," etc., etc., etc.
  - ○ The judges usually can figure out what the producer is trying to "tell" the audience simply by screening the show. What the judges want to know is who the target audience is, and what the target audience's action items are—what does the producer expect the target audience to do?
  - ○ Another irritating factor about this statement of objectives is the use of the word "tell." That's exactly what the show did: it told, and told, and told. Kinetic visuals were _irrelevant;_ almost all of the information was contained in the sound track. Noise! Ugh!
- _Gobbledygook:_ "To explain in simple and nontechnical terms the key ideas and benefits of object-oriented programming, a new programming technique used in artificial-intelligence applications. The main challenge of the program was to visualize highly abstract technical concepts by analogy with everyday life experiences."
  - ○ Frankly, I don't have a clue what the entrant is trying to communicate—nor am I much motivated to find out. The negative mind-set generated by such a prolix statement of

objectives is tough to overcome—and almost never is. Sometimes a statement of objectives is well written, succinct, precise, etc., yet is still inadequate—it proffers unattainable goals. The producer attempts too much and almost always fails. Two examples:

- *Unrealistic:* "This ten-minute video will teach aspiring medical students how to perform brain surgery." (I fabricated this one to illustrate my point, although it's not too far from some I've seen.)

This kind of show struggles to accomplish an impossible task. It usually deals with perceptual motor-skill development that requires precise actions performed in an exacting sequence. This type of skill is the most difficult to teach with linear motion-media even when the goals are realistic. However, such perceptual motor-skill training is ideal for interactive multimedia.

- *Too Ambitious:* "This film will recruit electronic engineers, give the annual report to stockholders, convince the buying public that our corporation is interested in the human condition and the environment, and persuade Congress that $750 is a fair price for a toilet seat." I also fabricated this statement. It's close, however.

Here's a classic case of moving heaven and earth for everybody, ready or not. Clearly, this sort of show tries to do too much for too many audiences, and does nothing for anyone—except, perhaps, put some to sleep. It's a prime example of preordained failure.

Writing a Strong Statement of Objectives

So far, we've emphasized the negative aspects of the statement of objectives. Let's explore the positive. A precise statement of objectives is composed of three parts.

- It begins with the overall intent of the show, best expressed with the infinitive verb form (one beginning with the word *to*). Examples are *to train, to motivate, to inform, to persuade, to sell,* etc.

- Next, there's a description of the target audience. This description must be as specific as possible. Some examples:
    - Eighteen-to-twenty-year-old male sailors who handle liquid

oxygen (LOX) on the flight deck of an aircraft carrier
- ° Executive visitors to our corporate office from other corporations or the federal government who are concerned with oil production quotas
- ° Retired senior citizens with fixed or limited incomes
- Lastly, a statement of what the audience is supposed to do, say, or think is needed—the <u>action clause</u>. Inherent in this clause should be the level of achievement expected and when it is expected. Some examples that tie into the audience examples listed above are:
  - ° To work safely, reducing LOX accidents to zero, both now and for the future, long-term
  - ° To press for legislation that will remove all federal controls on oil production in the upcoming congressional session
  - ° To stretch food dollars to realize a ten-percent gain in purchasing power now, and a thirty-percent gain within a year

Wrap. The first step in planning our show is to focus on clearly understanding its goal, defining the target audience, and articulating the action item. That's it.

Shelton's Pronunciamento

Ensure that our audience doesn't wonder, "What's this show about?"

Chapter 12

Who's Our Audience?

Audience Analysis

The most critical aspect of our pre-scripting analysis is determining exactly who our target audience are—whom are we attempting to influence? Typically, in linear media, our audience is a group and we seek insight into the sociological background of the individuals within the central part of the group. Accordingly, we encode messages in a *mise en scène* that is relevant to that central part of the audience—to engender empathy.

Before we get into the details of audience identification, it's prudent for us to review two audience communication theories.

Individual-Differences Theory

Even when our audience is homogeneous in a very narrow range—Navy LOX handlers, for example—all will not get the same message. Audience members are influenced differently and will react in diverse ways. The rationale for this non-uniform behavior is proposed by the individual-differences theory, which recognizes that an audience is composed of individuals who react to communication in their own ways rather than as duplicate automatons. Individual selectivity of exposure, perception, and retention are the salient points in this theory.[2] This theory is summarized in the maxim that no matter what message we intend to send, some in the audience will receive another.

Let's explore three salient points of the individual-differences theory:

- Exposure. People tend to migrate toward communications whose scope, tone, and messages they are in agreement with. They avoid communications of content they do not agree with.[3]

- <u>Perception</u>. People tend to construe, or misconstrue, persuasive communications according to their own concepts and attitudes.[4]
- <u>Retention</u>. Message retention is enhanced for those who are in general agreement with the treatment; less so for those who are not in agreement.[5]

Simply, different people select, distort, and interpret mass-communication messages in ways peculiar to their individual psyches—they get from the messages what they can and want. Yet, on average, the overall behavior of the group is predictable, and this is the basis for successful mass-media communication.

Sometimes, in extreme cases, a few individuals will act almost inversely to the behavior desired. On Wall Street, this effect is called the Contrary-Opinion Rule. Generally, with all other factors being nearly equal, this effect reflects problems with the individual rather than the show. It's the "there's-one-in-every-crowd" syndrome. Probably, we ought to discount such behavior.

Social-Categories Theory

The social-categories theory of mass communication proposes that people with similar social and psychological characteristics will be <u>influenced</u> and <u>behave</u> in much the same way when exposed to mass-communication media. A number of researchers have substantiated this theory to a high level of statistical significance.[1]

Information motion-media shows made for the general public or large heterogeneous audiences oftentimes are couched in general terms around a central theme. No matter the task for our mass audience shows, we'll fail to engender empathy in some members. We frame the information in a mode that patronizes the sophisticates and thwarts the uninformed proletariat. Nonetheless, some eminently successful, general-public motion-media shows are:

- *Battle of Britain* from the *Why We Fight* series by Frank Capra
- *Harlan County, USA* by Barbara Kopel

- *Night Mail* by Basil Wright
- *The Plow That Broke the Plains* by Pare Lorentz
- *Triumph des Willen* by Leni Riefenstahl

I categorize these shows as documentary rather than informational; however, they contain much information that is communicated exceptionally well, and they were instrumental in manipulating the attitudes of mass audiences.

Target Audience Profile

We use the social-categories theory to develop a target-audience profile to define the theoretical persons in our audience. The primary audience identification factors are:

- Demographic: age, gender, health and well-being, geographic location, and marital status
- Socioeconomic: education, income, occupation, social status, race or ethnic background, and religion
- Psychological: innate intelligence, sophistication, emotional profile, responsibility, character, industry, and experience with the subject

When we design our target audiences incorporating these factors, our shows are conducive to audience acceptance and understanding, and thus to communication.

Shelton's Pronunciamento
Know your audience.

Chapter 13

The Communication Analysis and Motion-Media Communication Plan©

To facilitate the development of communication analysis, I've included, at the conclusion of this chapter, my copyrighted form titled *Communication Analysis and Motion-Media Communication Plan,* with instructions. I've designed this form to codify the relevant information we need to know for topic research and script design. Faithful completion of this form augurs well for an accomplished show—all factors considered. My form is not "etched in stone." Modify it to meet the particulars of the communication problem at hand.

Communication Analysis

It is important to complete all the required information as accurately as possible to ensure a clear understanding of what the motion-media show is about, what is expected from whom, and when. Obviously, not all of the data required will be known after the first meeting with the client. To uncover all the pertinent factors, it probably will take several such meetings, interviews with the proposed audience, and other research.

Let's review the line-by-line items on the form. I suggest that you duplicate the form to follow coherently my discussion of these items.

- Proposed title. If the title is not finalized, put in a working title, or several proposed titles.
- Reason to produce this motion-media show. Pinpoint exactly why this show is being produced. What problem, function, or issue is it

intended to solve or alleviate? Is the problem real or imagined? What is the impact of the problem? In the long term and in the short term.

- Target audience. Define the target audience in terms as specific as possible. Some examples are "tenth-grade students studying chemistry," "the general public," "professional engineers who visit our organization," and "impoverished senior citizens."

- Purpose. The purpose or statement of communications objectives should be in three parts:

    1. Define the sponsor and/or producer's intentions. Over the years, I have found that this part ought to begin with the infinitive form of the verb, "to _____." Complete it with one of the following words (or another appropriate one): *arouse, canalize* (modify an existing attitude), *indoctrinate, influence, inform, motivate, orient, propagandize, report, train, teach,* or *sell.*

    2. Specify to whom the sponsor and/or producer's intentions are directed—the target audience, as I discussed in Chapter Twelve.

    3. State the action clause. It also ought to begin with an infinitive verb form. What is the target audience supposed to do, say, learn, understand, think, or feel—what is their action item? What level of achievement is expected and when it is expected?

- Target-audience profile. As carefully and completely as possible, describe the target audience in the three profile factors: Identification Factors, Motivation Factors, Psychology Factors. Obviously, it's almost impossible to develop an all-encompassing and near-precision target-audience profile. However, it's essential that the salient characteristics be identified and listed, thereby forming a comprehensive profile of the target audience. Again, see Chapter Twelve for details on the Target-Audience Profile.

- Secondary audience. Would other audiences benefit from the show? Who are they? It's important that the integrity of the motion-media show not be compromised for the benefit of secondary audiences.

- Essential elements of information (EEI). What information must be communicated to the target audience? In a linear show, the information ought to be broad-based notions or general guidelines. Three or four elements are optimum. If there are more than five, the show is probably too ambitious and has only a slim chance of success. However, in a multimedia show there can be more EEIs. We need be cautious and not engulf the audience members.

- Technical Quality Needed. Considering the difficulties inherent in achieving the communication goal, what is the minimum level of technical sophistication required to achieve this goal?

- Schedule. Pinpoint as accurately as possible the due dates for the cardinal phases of research, script design, production, and post production. With the client's approval, update the schedule to meet changing conditions.

- Filmic approach. The resolution of these elements defines what filmic techniques we use to produce our show—how all the artistic, psychological, and technical elements are integrated to form the final show—the grammar and syntax of the medium.

- Tenor. This is the overall personality of the show. Is it frank or subtle, serious or humorous, calm or excited, telling or asking, authoritative or dubitable? Perhaps it's some combination of these qualities.

- *Milieu.* This deals with the setting or environment of the show. Where and under what conditions does the scenario occur? Is the atmosphere real, fictional, or surreal? Dramatic, documentary, or *cinema vérité*?

- Characteristics. These describe the production techniques.

  1. Will actors or real employees be used, or animation, or some combination of live action and animation?

  2. Does photography include high-speed (slow-motion), time-lapse, micro- or macro-photography, special effects, or art and animation?

  3. Does sound include synchronous dialogue or voice-over narration, theater-of-the- mind, or stream of consciousness? Or is it some concoction of all these? What language is spoken? Stock or original music? Stock footage or slides?

- <u>Form</u>. This refers to the physical structure of the show. That is, the details of production: medium or media, length, gauge, black-and-white or color, digital or analog formats. Is the show complete or is it in several parts? Follow-on discussion? Handout material?

- <u>Communication surround</u>. Where will the audience view the show? What is the size of the screening site? What are the site's appointments and acoustics? Lighting? Ambient noise and light? Does the show require a leader or proctor to coach and monitor the audience? What qualifications are required? What is the availability of leaders or proctors?

- <u>Backup material and equipment</u>. What support material and equipment are needed for a successful screening/viewing? Consider such items as handouts, publications, photographs, charts, testing materials, models, chalk, blackboard, pointer, and microphone and speaker equipment.

- <u>Controlling factors</u>. These are usually client-imposed conditions: budget, schedule, distribution scheme (where, when, how), technical constraints, and information obsolescence.

- <u>Due date</u>. How much time is available to produce this show? When is the completed show due in the client's hands?

- <u>Serialized</u>. Is this show part of a series? How many others are in the series? Are they sequential? Do they stand alone? How much background information is needed in this show to maintain continuity?

- <u>Part of total communication package</u>. Are other media used to make up a total communication package? How is the show integrated into the total package? What is the nexus?

- <u>Changes or updates</u>. Will this show need to be changed or updated after its initial release? Visuals only? Sound only? How important are these changes/updates? How frequently must they be made?

- <u>Technical and political production considerations</u>. What concerns are anticipated during research, scripting, production, and distribution? Consider such factors as availability of key organization personnel, equipment, facilities, travel, locations, weather, contacts, and cinematography permits.

- <u>Hazards and safety considerations</u>. What hazards are likely to be encountered during production? What safety precautions need be taken? Special equipment? Clothing? Special insurance?

- <u>Client concerns</u>. Clients zealously strive to project and maintain a positive image. For some, it may be their most important asset. They are particularly sensitive to images projected or perceived in mass-communication media from which negative interpretations can easily be made that are hard to rectify.

- <u>Image projected</u>. What image does the client want the show to project? Is it to offset an existing negative image? What topics are sensitive? What are the pitfalls?

- <u>Company/organization policy</u>. What management position has the client taken on the topics covered in this show? What is the client's standard practice? Procedure? Rules?

- <u>Legal aspects</u>. What impact will this show have on any pending legal actions? Will it precipitate legal action? Are the statements made in the show defensible? Are they accurate? Do they affect union contracts or negotiations? Consumer groups? Copyrights?

- <u>Political impact.</u> What impact will the show have on other organizations? Governments? Employees? Consumer groups? Unions?

- <u>Proprietary information</u>. Is company proprietary or sensitive information included in the show? If so, what safeguards are there to protect this information? What would be the impact if such information were compromised?

- <u>Classified information</u>. Is classified government information included in the show? If so, exactly which visuals and audio are classified? What is the level of classification? What markings are required? How is distribution controlled? Crew to be cleared and for how long?

- <u>Budget</u>. What is the budget for this show? A motion-media show's budget should be separated into three distinct parts: scripting, production and post production, and distribution. Such a breakdown protects the client and producer, and ensures that we produce the best possible show for the resources allotted.

- <u>Media selection</u>. Nowadays, with technology rapidly developing, sometimes we forget that there are producing media and distribution media.

- <u>Producing medium</u>. What medium has the best potential to accomplish the communication goal set for this show? Consider all the factors discussed in relation to each medium's advantages and disadvantages (use the form below). Consider: what medium has the communication advantage—what medium engenders optimum communication in the most cost-effective way? Deciding which medium to select is a difficult task. For a detailed analysis of medium selection, I recommend consulting *Selecting Media for Instruction* by Robert M. Gange (Englewood Cliffs, N.J., Education Technology Publications, 1983).

- <u>Distribution media</u>. All factors considered, in what medium should this show be distributed to maximize communication in terms of quality, cost, timeliness, urgency, etc.? Do extraordinary situations or audiences require that this show be distributed in a secondary medium? What medium would best accomplish the secondary distribution goals?

- <u>Key Personnel</u>. List contact information for all important personnel involved in the production of this show. Such a detailed listing facilitates harmonious business dealings: Client, Technical Advisor, Approving Authority, Producer, and Script Designer.

Following is a blank form. You may duplicate and use this form as often as you like. However, please note on your plan that this copyrighted form is "Courtesy of Shelton Communications." See Appendix Three, page 205, for a completed form for the show *The Scarf: The Perennial Fashion Statement.*

Communication Analysis and Motion-Media Communication Plan ©

Today's Date: _____

Proposed Title:_____

Client/Sponsor: _____

Reason to produce this show: _____

Target audience:_____

Purpose: _____

Target audience profile

    Identification factors

        Demographic: _____

        Socioeconomic:_____

        Psychological: _____

    Motivation factors

        Anticipation: _____

        Importance of goal achievement: _____

        Urgency of communication: _____

        Information currency/obsolescence:_____

    Predisposition factors

        Sponsor: _____

        Communicator:_____

        Information: Interested. _____

        Medium: _____

Secondary audience. _____

Essential Elements of Information

1. _____
2. _____
3. _____
4. _____

Technical Quality Needed: _____

Interaction: _____

Schedule

Research: _____   Production:_____

Treatment: _____   Post production: _____

Script 1st draft:_____   Duplication:_____

Final script drat:_____   Distribution: _____

Filmic Approach

Tenor:_____

Milieu: _____

Characteristics:_____

Form: _____

Communication Surround

Audience size at each screening: _____

Frequency of screening: _____

Physical environment of the viewing site: Leader or proctor: _____

Projection/viewing equipment: _____

Proctor: _____

Power requirements: _____

Backup Material and Equipment: _____

Controlling Factors

    Due date: _____

    Serialized:_____

    Part of total communication package: _____

    Changes or updates: _____

    Technical and political production considerations: _____

    Hazards and safety considerations: _____

Client Concerns

    Image projected:_____

    Company/organization policy:_____

    Legal aspects: _____

    Political impact: _____

    Proprietary: _____

    Information:_____

    Classified information:_____

Budget

    Script/Storyboard: $ _____

    Production: $ _____

    Distribution: $ _____

Medium Selection

    Producing medium: _____

    Foreign language versions:_____

    Primary distribution medium:_____

    Secondary distribution medium: _____

Sponsor's Key Personnel

    Client/sponsor

        Organization:_____

        Names of contacts:_____

        Job title: _____

        Telephone number: _____

Address: _____

Other contact information: _____

Technical Advisor/Subject Matter Expert

Name: _____

Job title: _____

Department: _____

Telephone number: _____

Address: _____

Other contact information: _____

Client/Sponsor Approval Authority

Name: _____

Job title: _____

Organization: _____

Telephone number: _____

Address: _____

Show's Key Personnel

Show's Producer

Name: _____

Job title: _____

Organization: _____

Telephone number: _____

Address: _____

Other contact information: _____

Script Designer

Name: _____

Job title: _____

Organization: _____

Telephone number: _____

Address: _____

Other contact information: _____

Chapter 14

Filmic Design

To use motion-media to communicate effectively, we need to master and employ filmic design to optimize the plasticity of the media—the manipulation of time and space. Filmic design is a nebulous concept that can't be defined with precision. Nonetheless, I'll try.

> *Filmic design is the grammar and syntax of the medium— the filmic techniques we use to encode the messages into relevant kinetic visual and aural signals. In essence, filmic design is that concoction of psychological manipulation, technical prowess, and artistic achievement that we use to design and produce our shows. Each such concoction is unique, given the infinitely variable potential of filmic design—in essence, the grammar and syntax of motion media.*

Filmic-Design Techniques

Filmic design encompasses several broad (and perhaps overlapping) functions: filmic characteristics, milieu, tenor, approach, and execution. The specific items listed under each of these functions are some of the most common. Producers often combine several of these styles in a show for greater impact and audience involvement. Factors considered in determining filmic-design techniques are communications objectives, nature of the audience, budget, schedule, and preferences of the producer and client.

This list is not intended to be definitive. It's a general overview, and it's subjective.

Filmic Characteristics. These include the milieu, the physical elements that define the media. Genre. Setting. Arrangement. Form. The primary kinds of form are:
- Information
- Documentary (nonfiction)
- Narrative (dramatic)
- Enrichment (personal expression)
- Avant-garde (experimental)

In Chapter Seven, "The False Reality of Motion-Media," I discuss these genres in detail. (page 65)

Filmic Milieu. Refers to the style in which motion-media images are captured and presented. It's the environment, condition, or elements of the motion-media show. Some of the more popular are:
- Show-and-tell: traditional voice-over narration with picture and sound in coherence; archetype report; technical; information; education; etc.
- Narrative: dramatization or slice-of-life scenario
- Documentation: recording of uncontrolled actions, sporting or scientific events
- Historic past: actual or purported footage from the past
- Metaphor: scene is something other than it is, an analogy
- Objective past: past becomes present
- Past as memory: character's memory is the past

Filmic Tenor. Refers to the tone of the motion-media product—its mode of expression. The "feel" of it. Tenor affects the emotional response of audiences and their willing suspension of disbelief. The show may be:
- Authoritative or dubitable

- Calm or excited
- Frank or circumspect
- Serious or humorous
- Straightforward or subtle
- Telling or asking

Filmic Approach. Refers to the method of encoding information in the media; how the sight-and-sound images are recorded and presented. All shows noted as examples are cited in Appendix Two, starting on page 193.

- Cinematic continuity or discontinuity
  - Example of continuity is in *The Memphis Belle: A Story of a Flying Fortress* by Lt. Col. William Wyler, U.S. Army Air Force.
- *Cinema vérité* (literally "cinema truth"; sometimes dubbed "direct cinema"): unrehearsed recording of sight and sound of real people at work, play, etc.
  - *Harlan County, USA* by Barbara Kopple
- Flashback
- Kinestasis: extremely fast montage of still images, photographs, drawings, etc.; an individual shot may be on the screen for a very short time: perhaps only 1, 2, or 3 seconds, or even faster.
  - *American Time Capsule* by Chuck Braverman
- Players: professional actors or "real" people
  - Use of actors: for example, *My Father's Son* by Gerald T. Rogers
- Kino eye: the camera penetrates every detail of an unstaged scene and doesn't interfere with spontaneous actions.
  - *Regen (Rain)* by Joris Ivens
- Self-reflexive cinema (sometimes dubbed "self-referential"): an on-screen actor turns to the camera and speaks to the audience.
- Stream of consciousness: character's thoughts as voice-over narration or images
- Subjective camera: zero-degree camera angle—audience's eyes have the view of the camera lens; usually used as a close-up or extreme

close-up shot to focus the audiences' attention on some intricate detail of import. Subjective camera is especially effective in teaching perceptual motor skills.

  ° *Claymation* by Will Vinton and Susan Shadburne
- Talking head. An actor, expert, or downtrodden person faces the camera or some off-scene entity and talks. (See Chapter 19, page 161, for Hal Holbrook as Mark Twain.)
  ° *Face of Lincoln* by Wilber T. Blume and Dick Harbor, USC
- Theater of the mind: today's images with sounds of the past or future, or vice versa.
  ° *299 Foxtrot* by S. Martin Shelton

Filmic Execution. This refers to those cinematic and artistic procedures that we use to produce our shows. There are far too many such items to list and discuss here. Nonetheless, it's to our advantage to be technically proficient, to know equipment capabilities and drawbacks. We learn and hone technical prowess by hands-on experience and self-study. Many excellent publications abound on cinematic equipment and digital technology. Because technology is changing so rapidly, we ought to subscribe to several trade magazines that highlight these developments. One example is *American Cinematography*— the journal of the American Society of Cinematographers.

The filmic-design techniques I listed in this chapter are but a representative sample. With experience, we'll expand our palette to include many others.

Shelton's Pronunciamento
The importance of the message overwhelms all filmic considerations.

Part IV: Script Design

Chapter 15

An Information Motion-Media Writer Should Be a Script Designer

Traditional View

Frequently, I hear the notion that a competent writer (technical reports, press releases, novels, speeches, or articles) is equally adept at writing information motion-media scripts. It's the belief that "a writer is a writer is a writer."

It's true. Writers write—the written word is their primary communication medium. This is as it should be. Oftentimes, however, when written-word writers are tasked with writing information motion-media scripts, they tend to concentrate on the development of the oral commentary and write it as if it were to be read by a reading audience instead of spoken to a listening audience. By indirection, the writer has made the spoken word, narration, or dialogue the primary communication element. Thus, motion-media's most powerful communication element, kinetic visuals in related juxtapositions, is ill-used or negated.

Shelton's Pronunciamento
The commentary in a motion-media show is meant to be heard—not read.

Over the years, I've seen far too many instances in which entire script pages do not have a word, sketch, or symbol of any kind that deals with kinetic-visual development. Here are several examples quoted from the many that have been sent to me for review and approval.

- I see far too often the instruction "show it." (I wonder what "it" may be.)

- Here's a classic: "Add appropriate scenes to complement the narration." (Pray tell, where does one retrieve "appropriate scenes.")
- For a <u>narrative</u> scene dealing with drug addiction: "Use the right actors for this scene." The dialogue spilled over two and one-half pages. (Words fail me, as it were.)
- "The music ought to fit my words." (This was not a musical show.)

Shelton's Pronunciamento

Any writer who believes that narration or dialogue is preeminent in information motion-media communication doesn't belong in our profession.

Script-Designer Concept

I use the term *script designer* to describe the motion-media communicator who is known traditionally as the "scriptwriter" or the "film writer." "Script designer" is more than just a change in job title; it's a change in function. The script designer's task is to develop a viable scheme for solving communication problems with kinetic visuals supported by audio signals. It's a discipline in kinetic-visual communication. In effect, the script designer has the wherewithal to encode information into tangible filmic messages.

The script designer develops a storyboard that details the kinetic visual encoding of the show's messages. (See Appendix Three B for the storyboard of the show *The Scarf: The Perennial Fashion Statement.*)

Shelton's Pronunciamento

The information motion-media script is a visual design plan—<u>drawn</u> as a storyboard.

Chapter 16

Scripting the Information Motion-Media Show

Overwritten commentary is the primary fault in too many of our shows—too many words <u>telling</u>, and <u>telling</u> the audience too many things instead of <u>showing</u> them.

Shelton's Pronunciamento
We only use aural commentary to amplify or explain what the audience must know but cannot perceive from the visuals.

Writing per se is probably the least-important component in scripting an information motion-media show. Motion-media scripting is, in fact, kinetic-visual designing. "Don't write, draw," is an ideal maxim for us to follow in script designing—that is, don't *write* a script, *draw* a storyboard—no matter how primitive. I define an information motion-media script as consisting of the storyboard and the preproduction plan. (See page 206 for details of the preproduction plan.)

Our script—no matter how well conceived and detailed—should be mutable, in order to meet varying requirements, new perspectives, unforeseen production difficulties, and to encourage industry and take advantage of opportunities during production and postproduction.

Outline for Generic Information Motion-Media Scripts
There's no single way to design a script that will accomplish the communication goals set for our show. What I can do is suggest a generic script outline

that will fit many of our information motion-media communication tasks. It's a formula, modified in many ways that I've used over the years with some measure of success.

- Open the show with a "grabber" to get and hold the audience's attention and whet their appetites. My grabbers usually are fast-paced, dynamic, filmic montages of relevant scenes underscored with upbeat music. Minimal, if any, commentary. It's the "heads up."
- Set the stage by summarizing what the show is about, and state the communication goals.
- Communicate to your audience why this show is important to them. What advantage will the audience obtain? How will it affect them?
- Communicate the main points (EEIs) rationally, cogently, and empathetically. Such communication is accomplished through trenchant and dynamic visuals and graphic words in a pinpoint communication thrust.
- Tie concepts together in a memory chain linked together with smooth and relevant transitions.[1]
- Answer the primary questions expected from the audience. A pre-script test of a sample of the target audience helps define such questions.
- Use logic and persuasion to lead the audience to a conclusion (your communication goal) that is obvious and inevitable.
- Summarize the main points.
- Communicate the audience's action items and time frame for completion.
- End on a positive note, if appropriate, with a short filmic montage or another upbeat short sequence.
- To reinforce communication, consider providing the audience with take-home/office material.

In its simplest form, the basic outline is:
- Show 'em what you're going to show 'em.

- Show 'em.
- Show 'em what you showed 'em.

I'm not suggesting that we should repeat the information three times. Rather, we first introduce the topic: set the stage, as it were. Next, expand the information in appropriate detail. Last, summarize key points and state the audience's action items. What do we want them to do, say, or think as a result of having viewed the show?

Use this generic outline as a guide in your scripting ventures. It's flexible. Add to it, subtract, or revise it to fit the communication goals of your show. It's not the outline that counts. It's your thinking that counts—your ability to devise a filmic solution to the communication problem at hand. Often, the particular circumstances of each show dictate the appropriate structure. (In chapter 18, I discuss guidelines for writing narration and dialogue.)

Remember, *short is sweet*. For optimum communication, we design our motion-media shows with a pinpoint communication thrust and stripped of all the folderol. Audiences empathize easily with shows that are relevant, clear, and concise.

Storyboard

A script that engenders a high order of communication contains sketches or photographs and visual annotations, plus appropriate narration or dialogue. Such a script is a storyboard: a series of drawings, however polished or crude, of each planned scene, organized in a progressive continuity—the show's filmic structure.

For a complex animation scene, the script designers coordinate with the animator/computer artist for input regarding the technical complexity and the cost involved in executing the scene correctly.

Commentary

As I've been expounding in these past fifteen chapters, it is an indisputable fact that the communication effectiveness of an information motion-media product lies primarily in the kinetic visuals and secondarily in the commentary.

Recall that about 75 percent of the information in motion-media communication ought to be in the kinetic visuals and only about 25 percent in the commentary. Such a 75/25 ratio augurs well for long-term audience reception and retention of our messages. The 25 percent for commentary appears to be the effective maximum limit and is applicable only when the commentary directly complements the kinetic visuals, i.e., "show and tell." The two should cohere. Commentary that's inconsistent with the visuals interferes with communication, introduces noise, and reduces a show's effectiveness.

Shelton's Pronunciamento
The information encoded in the kinetic visuals and in commentary must be in coherence.

Effective commentary must be brief, trenchant, and not intrude on or upstage the visuals. Commentary may be a curt phrase, an exclamation, a flashing thought, a complete sentence, or any manner of well-informed utterance. It may be narrative, meditative thought, rhyming poetry, or blank verse. No matter what its content or form, commentary's exclusive function is to amplify and explain what the audience cannot perceive from the visuals yet must know for complete understanding. What is critical is that the commentary sounds right. How does it sound? Is it natural to the scene?

Shelton's Pronunciamento
Commentary is first, foremost, and always an ancillary function to the kinetic visuals.

Ideally, scripts are completed and approved before production begins. Often, however, it is not possible to develop the script in detail before cinematography or videography starts. This situation would apply, for instance, in those information motion-media shows that are composed mostly of stock footage, or instrumentation (high-speed), test-and-evaluation, medical or scientific footage, or that of news events or other activities that cannot be preplanned, controlled, or staged. Such lack of pre-scripting would also apply in scenes that show military, technical, or other specialists speaking their own jargon and doing their own jobs in their own particular ways. In such cases, we can't predict with any certainty the exact nature of the recorded images. Accordingly, such a show's final design depends on the cogent editing of the 'what happened' scenes, and in developing relevant narration.

Even when we've preplanned in detail the commentary for tightly controlled situations, flexibility is key. To meet changing circumstances during production, commentary must be adapted to conform to the cinematography.

Information Film/Video Archives

Part of our job as script designers is to know and evaluate the cost and expected results of the scenes we develop. For instance, if we were to go to all the worldwide locations to film scenes that show the various types of scarves called for in our storyboard in Appendix Three B, *The Scarf: The Perennial Fashion Statement,* our production budget would skyrocket and we'd be out of business. The answer: use drawings or animation, or stock footage. Quite literally, there are millions and millions of feet of scenes of all types and captions available for our use in this country and throughout the world. For example, stock footage is available through

- Stock footage libraries
- Television networks (broadcast and cable) and local stations
- Film/video production houses
- Corporations
- Universities
- Motion-picture organizations
- Museums and libraries
- Government offices (the Library of Congress, for example)

Sometimes the stock footage is free; most of the time a fee applies. The cost usually is based on the amount of footage used in the final show and on how the show is to be distributed. Unless we're familiar with an organization's stock-footage catalog, we should hire a researcher to find the scenes we need.

On a per-hour basis, the cost of a researcher may seem high, but using an experienced researcher who knows the resources is a worthwhile investment. Costs also include paying for duplication and shipping.

Also found in stock libraries are still photographs, music, and sound effects. It's imperative that we always get written permission and a license agreement anytime we use stock material. Such is especially the case with music; otherwise, we may well end up as defendants in court.

Pace

Today's audiences are accustomed to thirty-second and fifteen-second television commercials with a rapid-fire barrage of hard-sell messages that inundate the viewer. If this pace is used to excess in information motion-media, the audience will "escape" mentally. It's important to give our audience reflective time to assimilate and integrate newly presented information. We need to vary the pace of our shows to get and maintain audience interest.

Script formats

I've used several script formats in my career. Which format to use depends primarily on the show's *mise_en scène*. Traditionally, we use the split-page for information motion-media scripts that call for voice-over narration—the anonymous, authoritative, and unseen voice.

We use the teleplay format for slice-of-life and real-life dramatization scenarios in which actors speak scripted lines in synchronous sound. The teleplay format includes several specific types: "business teleplay," "corporate teleplay," and "information teleplay." These types of information motion-media shows mimic the style of narrative motion-picture films and television programs. Nowadays, script designers are using the teleplay format more frequently because more and more of our scripts are set in the talking head or narrative *mise en scènes* (unfortunately).

No matter which format we use, what's important is that the script be organized clearly, graphically lucid (with a storyboard), and vividly unambiguous.

To illustrate the primary script formats, I've included four samples in the appendices:

- Appendix Three: Split-page and storyboard. The script for the show *The Scarf: The Perennial Fashion Statement*. Please note that there is <u>no narration or dialogue</u> in this script.
- Appendix Four: Teleplay format. A few pages from the script for the show *Gambling Addiction and the Family*.
- Appendix Five: Split-page with voice-over narration. A sample of the script *Desert Stewardship*.
- Appendix Six: Split-page and teleplay combination. The opening sequence of the show *Pacific Frontier*.

Conclusion

A well-designed information motion-media show portends success. Communication goals are achieved, and resource expenditure is optimal. In such a scenario, clients and film designer benefit, and the audience does not try to escape.

Chapter 17

Marty's Contrary Principles of Script Design

Unfortunately, the scripts for most information motion-media shows are inept. Mimicking commercial television or the "movies," they employ dialogue or narration to carry the bulk of the information. The script may read well on paper, but it doesn't enable much communication when translated to a kinetic motion-media sight-and-sound show.

To illustrate, I've codified the most common scriptwriting problems that comprise my "Contrary Principles." From this point on, I use the term "scriptwriter" pejoratively. The following list is not in any particular order, nor do I pretend that it is complete. (Read on, please keep your laughter in check.)

Overwritten. Narration/dialogue tells the audience everything they'd ever want to know about the topic. The audience is beclouded in irrelevance. They cannot perceive the essential elements of information or what is expected of them. Rather than inundating the audience with drivel, concentrate on the kinetic visuals and make the verbiage sharp and concise, lean and mean. In motion-media, silence is golden.

One more comment on overwriting. Some years ago, when I was the manager of the motion-media competition for a major communication society, we received a video for a target audience of deaf people. As incredible as it sounds, the talking head was the video's entire filmic *milieu*. We can't assume that the viewers were lip readers.

Punctilious. To ensure 100-percent correct grammar, form, and syntax, the scriptwriter writes narration or dialogue to be read rather than heard. When it's heard, it sounds stilted and academic. Scripts are not an extension of literature. It's okay to develop narration/dialogue in non-sentences, to use idioms and slang, and in general to be less formal than when writing for a reading audience. Conversational style is appropriate for a listening audience.

Discordant. Dichotomy between visuals and audio: two disparate messages are sent simultaneously, each element going its own independent way. The audience is confused by the mixed signals. Discord is common in far too many television commercials. In effective motion-media, there is *absolute coherence* between visuals and audio. Audio reinforces the visuals to form a synergism that greatly enhances communication—show and tell.

Tautological. Visuals and narration are redundant. Words tell the audience exactly what they can see on the screen. Instead, words must be used only to tell the audience what they cannot perceive from the kinetic visuals yet must know for complete understanding.

Radio-ish. Vapid and irrelevant visuals are used as a mechanical device to use up screen time while the bulk of the information is transmitted in the narration. The visuals do not pertain to the points being discussed, but they don't directly interfere either. Consciously or subconsciously, the audience tries to interpret the visuals (in what ought to be the primary communication source in the sight-and-sound media) but cannot. They become confused and distracted, and the result is poor communication. Question: Why is radio drama such a powerful communication medium? Answer: It's because we, the listening audience, create our own perfect and powerful visuals. We're not confused with irrelevant visuals that distract our concentration.

Malapropos. Narration is not in a style and tone appropriate to the target audience. For example:

- Passive. Passive voice dulls the audience's mind. Active voice in present tense stimulates and involves the audience—empathy. And stimulation is the essential ingredient in communication.
- Pedantic. Pompous language and scenarios are used to impress the audience or client with how learned the scriptwriter is.
- Telling. Telling style tends to alienate many audiences. Use carefully.
- Over and Under. Narration is cast in a form that is either orotund or patronizing to the audience. It misses the understanding and needs of the audience.

Desultory. The show rambles and lacks structure, and audiences cannot follow it. I firmly believe a motion-media show should have a beginning, middle, and end. The sequences ought to be linked in a chain of associated ideas that build to a logical conclusion. Such products, however, do not necessarily have to have linear development.

Farraginous. A hodgepodge; a confused mixture of too many goals for too many audiences. The show is too ambitious and lacks a critical focus. It's like an ice-cream sundae made from a dash of dozens of flavors. It tastes okay, but what flavor is it? What's the message the target audience is supposed to get from the show?

Vacuous. In an attempt not to offend anyone, the show communicates nothing to anybody. It would be better if these sorts of shows were never produced. However, they are politically correct and engender "feel-goodism."

Self-Indulgent. "Creativity" is used for creativity's sake and to massage the scriptwriter's ego. "Here's my chance to write a 'great' movie. I'll be 'creative.' I'll entertain the audience with production value, actors, locations, digital effects, and lots of razzle-dazzle, and in stereophonic sound." Wonderful! Hope you had a great time at the client's expense. But did the show accomplish the communication goal set for it? And was it worth the client's money and resources? Probably not! As Cap Palmer says, "The only conceptual kinship between a good [informational] film and a 'movie' is the accident of being packaged on long narrow strips of cellulose acetate through which a beam of light shines."

Gimmicky. Overabundance or misapplication of novel ideas, styles, and devices that call attention to themselves and detract from communication. Gimmicks are noise in the communication process.

Monotonous. The show lacks variety in rhythm and accent. It has one speed from beginning to end, no "dramatic" sense or development. Since motion-media is plastic, the effective script designer manipulates time and space to arouse the audience's emotions in order to engender commitment and empathy.

Rough. No transitions to lead the audience smoothly from one concept to the next—nothing to integrate the scenes. Today, many scriptwriters and producers have the notion that "audiences are TV-smart. Hit 'em on the head with raw information as fast as you can." From what I've seen in such shows, I'm convinced these folks are rationalizing. They won't (or can't) put out the initiative, industry, and imagination needed to make an all-around professional show.

Inept. Just plain incompetence in all facets of scripting and production. The scriptwriter doesn't know the syntax and grammar of filmic design, is not familiar with techniques or processes, and doesn't comprehend the scope of the resources required to execute these techniques successfully. Thus, the scriptwriter fails to use the medium effectively by either under-utilizing or over-utilizing techniques.

Stilted Scenarios. In an attempt to be "creative" and to make a narrative "movie," some scriptwriters couch the essential elements of information in a dramaturgical scenario laced with trite dialogue. Such scenarios are hackneyed and implausible, and are best avoided. The fact that the dialogue carries the bulk of the communication burden should give us a clue to the inherent problem in these scenarios.

Below, I've listed some particularly egregious stilted scenarios that are particularly ineffective and insulting. These scenarios are generic, and there are many variations within each.

- Man on the Street. In "spontaneous" utterance, random characters extol the virtues and evils of whatever point the show is trying to make. Almost no one believes that such dialogue is unrehearsed and undirected, and credibility is seriously eroded by such chicanery.
- Newscast. Contrived to a crippling fault are scenarios of a "newscaster" faking a broadcast with a news bulletin; these always fail. Sophisticated audiences recognize the flimflam instantly and are insulted, guaranteeing communication failure.
- Interview. Under the guise of a no-holds-barred interview, some "big-time journalist" questions the chief executive officer to get the "real" facts. Sure! The journalist asks the obligatory, scripted, arranged-in-advance, tell-me-what-we-both-already-know question, which is followed by the dittoed answer. It's patent tomfoolery, insulting, and ineffective.
- Old Guy Talking to the Young Guy. By far, this is the most popular and notorious stilted scenario. It's the scenario that I abhor with the greatest vigor! The basic premise of this scenario is that the sage (old

guy) tells the novice (young guy), in ping-pong question-and-answer routine, why to buy it or how to do it. "It" can be anything from:

- Coffee: "Madam, here's how to save your marriage with a better cup of our coffee."
- Automobiles: "Sir, here's how to inveigle that attractive female."
- Home mortgages: "Folks, here's your path to upward mobility, even if you're fiscally overextended."
- Field-stripping an M-16 rifle: "Listen up, recruits, your rifle is your best friend."

I wholeheartedly resent the fact that if I'm to get information from this back and forth badinage, I must eavesdrop on the actors' conversation to get the information I need. These obnoxious stilted scenarios dominate radio and television commercials. Ugh!

Talking Head. Our information motion-media shows are laced with talking heads that talk, talk, and talk. Ensconced in an appropriate setting decorated with pertinent trappings (CEO's office, perhaps), the talking head tells us why the Acme Toxic Waste Disposal Company treasures people and cares for the environment—and tells, and tells, and tells! At best, such a scenario is a photographically recorded lecture that almost no one cares a hoot about. Novices should reject all temptation to use this scenario. (Nonetheless, see Chapter 19, The Talking Head, starting on page 151.)

Parody. Frequently done, seldom successful, and always a self-evident reflection of the scriptwriter's limited imagination is parody. The scriptwriter tries to imitate the style of a well-known artist or the form of a popular television show or motion picture in a feeble attempt at humor or a real-life dramatization. Audiences feel cheated and patronized at such second-rate imitations of the real thing. "Don't we deserve, at least, an original idea?" they might rightly wonder. This sort of audience perception does not engender much communication, at least not the kind intended by the scriptwriter and client.

Real-Life Dramatization. In an outlandish photoplay, some downtrodden character's problem, emotional, social, or financial, is solved by the character doing what's "required"—usually dictated by family, friends, or coworkers. For instance, "Your nose warts disappear when you start using Acme Nose Wart Nostrum." Artificial plots, tediously developed in such scenarios, do not achieve communication.

Slice of Life. A hard-working, good-natured bumpkin is in deep trouble because of lack of judgment, insight, or knowledge. Someone with a special interest in the bumpkin sets him or her right by dispensing the missing insight or knowledge. The problem is solved, and all live happily ever after. Who's kidding whom? With today's attuned audiences, such a jejune scenario will elicit scorn and giggles.

Admittedly, I've skewed the thrust of my arguments here to fit my own predilections. I'll also reluctantly admit that, in some instances, under certain circumstances, perhaps an exceptionally well-crafted dramatic scenario might work just fine. However, I recommend that beginners and long-time professionals always avoid stilted scenarios. When a narrative show is required, I suggest that only an experienced script designer skilled in dramaturgy develop such a script.

The only proven way to improve your script-designing skills is to visualize, visualize, visualize, then draw, draw, draw your storyboard, no matter how crude. Finally, write a few lines of commentary—if absolutely necessary.

Shelton's Pronunciamento
Successful scripts and shows are *simple, short, and straightforward*.

Summary
We do not write scripts. We <u>design</u> scripts. We use filmic design to develop the kinetic-visual narrative. I recommend that scripts be critiqued by your peers and those whom you respect. If you're super-serious about this profession, pay for such critiques from an honest broker, someone who is recognized by our profession.

Gadzooks!
In the following chapter I'm "gonna" talk briefly about writing. Don't faint.

Chapter 18

Guidelines for Writing Narration and Dialogue

No matter how hard you've tried, you can't encode all the information in the kinetic visuals. Your storyboard is brilliant. The visuals you've planned are cogent, dynamic, and psychologically relevant, but they just don't complete the communication task. After lots of soul-searching, you're forced to admit that you've got to use commentary to tell the audience what they can't perceive from the visuals: voice-over narration, dialogue, or a talking head. What do you write? How do you write it?

In the following sections, I suggest some guidelines that'll help you develop commentary that contributes to communication. I've listed these guidelines in their general order of importance.

General Guidelines

Ensure that the narration and the kinetic visuals reinforce and complement each other—that they are in coherence. Ensure that the narration does not send one message, the visuals another. When the visuals and narration are in coherence, they create a dual media synergism that significantly enhances the communication potency of motion-media. The show is harmonious and effective. (See Appendix Five for the sample script "Desert Stewardship.")

- Write narration and dialogue as if every word you write costs you two hours of pay. Too many words telling the audience too much information pound the audience into a stupefying, catatonic state. The eye is our primary sensor. The eye receives information at a significantly faster rate than does the ear.

- All factors being equal, voice-over narration is the voice of authority for most audiences. The anonymous narrator speaks from "on high"—presuming that the narrator is a first-class professional. Direct the narrator to use the right mix of vocal tone and pace to emphasize the importance or urgency of the messages.

- Narration and dialogue are heard, not read. A conversational style is appropriate for the listening audience. Write for the ear, not the eye. A good test is to have someone read the narration and dialogue to you. How does it sound?

- Sometimes narration can be blank verse, as in the "Voices of Commerce" sequence in the *Song of Ceylon*. In the last scene of *Night Mail*, the narration is in rhyming couplets spoken in the tempo of the clickety-clack of the fast-moving train as it approaches Glasgow. Use such techniques with caution.

- Don't write narration that tells the audience what they can see for themselves in the visuals, or just to keep the narrator talking.

- The maximum amount of narration that's prudent to have in a film or video is about one-third of the total screen time, excluding titles and credits. (And I fervently hope that there is a lot less in your shows.) Should this 1-to-3 ratio be exceeded, the show begins to have major problems in optimizing communication. There's too much information in the narration.

Specific Guidelines

Here's a potpourri of specific guidelines to help ensure that your commentary is as excellent as it can be. Obviously, no list such as this can be complete. But do with it as you will: add, modify, or (I pray not) delete from it.

- Narration should not be noticed. It must be natural to the scene and natural in tone and syntax. Narration that calls attention to itself usually becomes noise in the communication process.

- Narration and dialogue must be written in a tone appropriate to the particular audience. Talking up or down to audiences ensures communication breakdown.

- Use dynamic, active-voice verbs in the present tense. Such verbs are stimulating and have an involving effect that maintains the audience's interest.
- Avoid passive voice, which dulls the audience's mind. Sometimes, however, the passive voice is appropriate, such as when the emphasis is on the object rather than the doer. For example, "Airplanes are made of aluminum," or "He was fired!"
- Narration doesn't necessarily have to have coherence and continuity in itself. (For example, in my film *299 Foxtrot*.) Filmic coherence should come from the kinetic visuals through montage editing and manipulation of the plasticity of the medium—filmic design.
- Write the narration in a natural, conversational style. Be careful of tongue-twisters.
- Short sentences and well-formed utterances are best. Concise, direct, easy-to-understand words and phrases communicate effectively.
- Include only one thought per well-formed utterance.
- Avoid pronouns whenever possible. In aural communication, antecedents get lost quickly and frequently.
- Avoid adjectives and adverbs that reflect your personal interpretation—for example, "beautiful," "quickly," "azure blue." Audiences make their own judgments and may resent your imposition.
- Use colloquialisms with caution.
- Be personal in your writing. In many instances, it's better to use personal pronouns such as "we" and "you" to establish a warm and friendly ambience. Be cautious, however: don't overdo it or you'll become smarmy. It's okay to use contractions from time to time.
- Use everyday words for your particular audience, and use them simply. Be natural.
- Use precise language. Say exactly what you mean and mean exactly what you say.
- Use simple, unambiguous words.
- Use concrete nouns and action verbs.
- Keep the nouns next to their verbs and follow with the object. Such

a verbal arrangement in oral communication optimizes communication. A natural order and uncomplicated construction are basic to
conversation.

- Develop vocal variety, rhythm, and pattern in the <u>sound</u> of the
  commentary.
- Repetition reinforces communication.
- Write instruction in the positive ("Do this") for maximum impact.
  Use "Remember" instead of "Don't forget," for instance.
- Introduce new information in context with old information. In some
  instances, it's best to start a sentence with a bit of old information
  and use a connector to end with the new, important information.
- In some "show and tell" sequences, presenting information via a
  "laundry list" is appropriate. For instance, "The veterinarian inocu-
  lated poodles, terriers, setters, and hounds." Ensure that the narrator
  puts a period after each of the nouns.
- Keep construction parallel for maximum impact.
- Avoid imperatives. Telling tends to alienate many audiences.
- Use acronyms only when you're positive the audience will under-
  stand their full meaning. Tread prudently.
- If you must use technical terms, use them—but use them sparingly.
  Consider using a superimposed title of the technical term near the
  bottom of the frame to reinforce communication.
- Abstract ideas are communicated effectively through trenchant visu-
  als (animation, perhaps) and graphic words.
- If you must use foreign-language names or phrases, give the narrator
  a break and spell them out phonetically in parentheses in your script.
  Superimpose the name or phrase near the bottom of the frame.
- Use alliteration, simile, and metaphor with caution. Often these rhe-
  torical devices are best used in writing that is to be read, not spoken.
- Use idioms, slang, clichés, etc., with extreme caution. I've found over
  the years that such sayings contribute little to communication.
- Round off numbers, if appropriate. Reinforce the communication
  with a superimposed title of that number.

- When dealing with the size of things, make comparisons with objects familiar to the audience. For example, "about the size of a dime." Better yet, show a dime next to the object and skip the narration. In *299 Foxtrot* we put an automobile key next to a bullet hole in the aircraft's fuselage.
- Be scrupulously honest. Any false comment or perceived attempt to mislead the audience ensures communication failure. The entire show is discredited.
- Avoid superlatives, such as "greatest," "bluest," "most."
- Avoid, if possible, comments that date the production. (Ditto visuals.)
- Avoid starting too many comments with words that begin with "th." Such writing sounds stilted and lacks variety and rhythm.
- Avoid overuse of "you" and "your." A few of these two words are appropriate.
- "Gobbledygook obfuscates ratiocination (as it were)."

Application of these guidelines with professional élan is the key to writing commentary that enhances communication in motion-media. Admittedly, there are exceptions. In that rare instance when you believe that you've got an exception, proceed with caution.

The following is a sample of superior voice-over narration. It's about one-half of the total narration of the award-winning seven-minute film *Vision Lab*—a sales film for a (then) new computer graphics system. Through use of a highly stylized format, we see the system in operation, demonstrating its capabilities. William (Bill) Mauger, president of Sunbreak Productions, was the executive producer of *Vision Lab*.

"Vision Lab"
The means to an end.
The end. Limitless transformation.
Imagination.
The means.
Vision Lab.

Non-dedicated distributed workstation image processing.
Software. A high-resolution monitor . . .
Vision Lab. Every function you imagined.
Archive. The storage and retrieval of images on disc.
Digitize. To capture an image 30 times a second or to freeze a single frame.
Analyze. Zoom. Pan. Region of interest.
Pixel coordinates and values. Plot pixel values.
Filter. Sharpening. Smoothing. Convolute an entire image.
Erode and dilate grayscale.
Image edit. Paint. Airbrush. Erase.
Cut. Paste. ..."
(Quoted with permission Sunbreak Productions)

Notice that the narration lacks continuity, detail, and formal grammatical structure. Also, please note that the narration is sparse. Mauger optimizes communication in this film by couching the vast majority of the information in relevant and dynamic visuals. He uses the narration only to reinforce the visuals. This film is a classic "show and tell," having the visuals and narration in near-perfect coherence.

Unfortunately, on the printed page, you can't see this film. Nonetheless, you ought to see this show to fully comprehend how Bill Mauger uses the syntax of our media to communicate commanding messages, using kinetic visuals and graphic words in harmonious coherence to form a powerful synergistic communication tool.

I'll conclude this chapter on writing narration with a comment from a student who attended my script-design workshop a few years ago. She was a senior technical writer/editor working for a large organization that has no in-house motion-media capability. She was tasked with "writing" a video script on a technical subject. Here are her comments: "Script design *works!* I find that the thing that works for me in writing narration is to 'hear' it in my head as I go—then I go back and pronounce the words. This inevitably leads to reducing syllables and eliminating words. Then I'm visualizing how the words and the images will be, as an audience member. The customer was very pleased."[1]

Chapter 19

The Talking Head

The talking head can be, and all too often is, the death knell of effective communication in motion media. We're inundated with talking heads in our kinetic sight-and-sound media. We *hear* everybody from the chief executive officer to the local politico, the television news-talker, "the standup" on the evening news, actors bantering in television commercials, or the man or woman on the street telling us all about it (whatever "it" may be).

The talking-head scenario is popular, quick, and cheap. Admittedly, the talking head does effect some communication, in varying degrees, to some audiences. But it's tedious, unfilmic, and woefully inefficient in communicating ideas. More often than not, the talking-head scenario reflects the script designer's lack of cinematic skill and filmic imagination, or just plain laziness.

In particular, I'm seriously disappointed in several cable news channels. The hosts ask their guest a series of important questions. And sure enough, when the guest answers, the producer begins to show stock footage of the topic, which may or may not be relevant, in a split-screen. Now, we have two or three (or more) images simultaneously on the screen. Where do we concentrate? This hodgepodge of visual imagery and commentary destroys effective communication. Skip the stock footage, and let's see the guest answering as a talking head. (Indeed, I said it!) At least the audience would know where to focus their viewing.

Partial Communication Power

As a communication tool, the talking head functions only minimally because the kinetic visuals of the "head" contain only minimal information. It is the oral comments from the "head" that contain the messages. Clearly, we're not utilizing motion-media's full communication potential, but, I reckon, some communication is better than no communication. Please recall that optimum communication in motion media is achieved in shows in which 70 to 80 percent of the information is contained in the kinetic visuals.

It takes me a long time to *tell* you what I want to communicate. But I can *show* it to you much more quickly and much more effectively.

Here's another way to illustrate why the talking-head scenario is so inefficient. Let's suppose our task is to drive from Los Angeles to Las Vegas as effectively and efficiently as possible. We have several options. We can drive our finely tuned, eight-cylinder roadster at 70 mph on the interstate highway. Or we can drive Honest Harold's used jalopy that runs on only three cylinders, and take unpaved, desert back roads. We'll get to Las Vegas either way. But when will we arrive? How much effort did we expend? More important, what's our frame of mind after we arrive? Our eight-cylinder roadster is a finely tuned filmic-design show. The jalopy is the talking-head show.

I'll admit that there is some visual information in the talking-head scenario. As we *see* such persons talking, their affect projects through the screen; we may discern their authority, intellect, demeanor, and sincerity (or we can be fooled). The setting and background help set the ambiance of the scene. In addition, other visual clues contribute to the communication process—a person's dynamics, delivery, dress, accessories, grooming, etc. But on the whole, such shows dodder along to communication oblivion.

Filmic the Talking Head Is Not

Far more times than I care to recall, I've seen talking heads in shows that reflect the worst in filmic communication. One classic example is the CEO sitting (hiding) behind the desk, and droning on and on, *telling* and *telling* the audience what they generally don't care a hoot about and don't need to know.

Shelton's Pronunciamento
Rapid-fire babble couched in gobbledygook pounds the audience into stupefying catatonia.

The Talking Head Communicates (Gadzooks! Did I say that? Where's the straight jacket?)

I reluctantly admit that in certain cases, the talking head scenario may be appropriate when *critical information* must be *transmitted urgently* and economically to concerned audiences who may be in different locations, and when prompt audience reaction is required.

Nowadays, the talking-head *mise en scène* usually is distributed via real-time communication technologies: the Internet, television, and other contemporary devices.

Accordingly, let's explore the talking head as the production medium for distribution to mass audiences.

Often, it's prudent to have some communication <u>now</u> even though it's couched in a marginal filmic *milieu* rather than to delay communication in favor of completing a highly efficient filmic show later. In the latter case, the value of the information will probably have expired or the required audience action be delayed and, so, it will be of minimal or no value.

The validity of the talking-head show is a function of the speaker's authority, credentials, and ability to deliver messages with conviction and verve. Use of relevant props, graphics, still photography, and stock footage/tape will make the show more filmic and, thus, enhance its communication effectiveness.

The degree of communication achieved is related directly to the intensity of the audience's *need* for the information, the *urgency* of the audience's action, and the *currency* of the message (when does the value of the information perish?). (See Shelton's Communication Theory in Chapter 3, page 23.)

No matter how dull or inept a motion-media presentation may be, communication will occur if the audience has a critical need for the information and must act quickly. For example, a television news announcer interrupts his spiel to warn, "A tsunami will engulf our city in ten minutes." Then, the announcer fools around with papers and continues, "Our sponsors' messages follow." *Egad! I hope not!*

Here are some instances when the talking head may be appropriate and effective:

- CEO announcing that tomorrow the plant will close permanently and all employees are dismissed; printed details to follow shortly
- Renowned scientist lecturing on the latest unified-field theory to a group of graduate students majoring in physics; formal paper to be published in the Acme Scientific Journal next month
- Evangelist preaching the evils of sin to the sinners; the book of divine inspiration is your guide
- Television newscaster describing little green men emerging from their flying saucer, which has landed on the White House lawn; video footage to follow

Though these shows aren't cinematic masterpieces, they'll accomplish most of the communication task.

Personal Approach

Particularly vexing is that nowadays it's common practice to have the talking head look slightly off-camera. That is, the head looks at someone we can't see—a person who is out of the camera's view (perhaps an imaginary interviewer). It's dishonest! If someone is going to talk to me, I want that person to look me directly in the eye and tell me forthrightly what's on his or her mind. I resent having to eavesdrop on a conversation to find out what's going on.

When the talking head looks directly at us, with strong eye contact, the communication is personal. It's one-on-one. We get a perceptive insight into the personality and character of the person talking to us. We can evaluate this person and make conclusions. Is this someone I can trust? Is the message true? Relevant? Do I care?

The personal approach often is preferred for the dissemination of "bad news." For example, if the CEO has to announce the closing of a plant to the affected employees, the talking-head *mise en scène* is appropriate. By couching the message in a personal, compassionate tone, the CEO may be able to soften the bad news. The personal approach cuts through the layers of interpretation that are inevitable. Jobs are being eliminated. Our audience becomes deeply involved and strongly identifies with what's happening.

Here's about the only directorial advice you'll get in this entire book. If you *must* use a talking head, always have the person look directly into the camera lens. And position the lens at that person's eye level.

Visual Ambience

It's the visual ambience of the *milieu*—the setting in which the talking head speaks—that strengthens communication by establishing the mood, providing visual clues, lending credence, and piquing audience interest. Included are such visual elements as the location and the background. It's important that the setting be an integral part of the story. The *mise en scène* or *milieu* is the place where the talking head works, lives, or does whatever the show is about.

To increase further the communication potential of our talking-head show, we can enrich its visual ambience by using filmic design to incorporate all manner of pictorial elements—kinetic, still, or any manner of visual aids—props, models, uncomplicated charts or graphs, titles, artwork, still photographs, or stock footage. Perhaps original footage can be recorded quickly to cover key points. These visual scenes can be used as cutaways from the talking head to show what's being said. It's the classic "show and tell" coherence that makes effective communication.

Shelton's Pronunciamento
Motion-media shows should be visually filmic.

One excellent example of a talking-head presentation that capitalizes on a host of visual elements in the natural environment was the Public Broadcasting System's series *Ancient Lives,* produced by station WTTW in Chicago in 1984. (It matters not when a show was produced. What matters are the filmic techniques.) The series is hosted by John Romer, British archaeologist, historian, and storyteller extraordinaire.

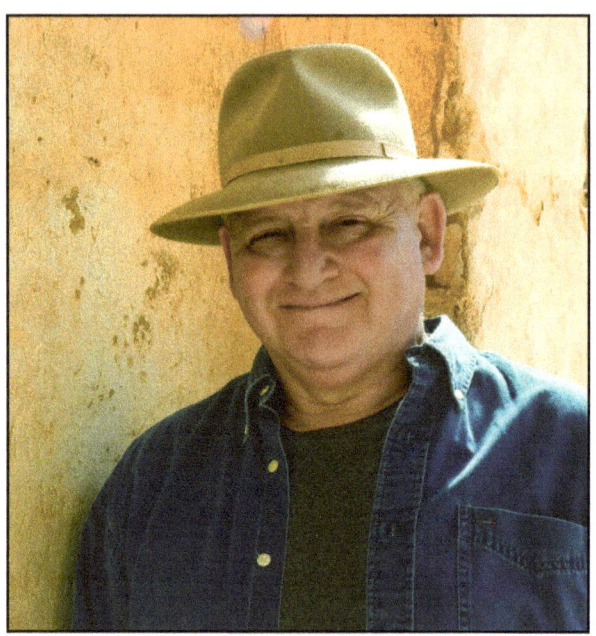

Figure 24. John Romer
Courtesy of WTTW, Boston

In one episode entitled "A Village of Craftsmen," Romer escorts us on a tour of ancient Egypt. His goal is to share the history of the times, including a sociological perspective on Egyptian society. Romer is on camera only about 40 percent of the time, weaving his story around a host of strong visual scenes. He interprets hieroglyphs, shown in dramatic close-up shots, and takes us to locations in the story such as Thebes, the Valley of the Kings, and Tutankhamen's tomb, and to the Cairo Museum, where we see artifacts, works of art, mummies, and other antiquities, each exactly fitting the points he makes.

Aside from its technical excellence and inherent subject interest, several factors combine to make this talking-head presentation a highly effective communication tool. For instance:

- The background is natural. It does not interfere with the communication. (It's not noise.)

- Romer gives scale to the environment and to the props.
- Relevant cutaway visuals are used extensively.
- Romer convinces us that he is an expert Egyptologist. We tend to believe him; he's established his credentials.
- Romer looks directly into the camera's lens—that is, he looks us in the eye.
- Romer is a consummate storyteller. His personality projects through the lens, giving the series a personal flavor. He's empathetic.

Identification

Fundamentally, such a talking head is a storyteller. Our audience is on location with the storyteller, and they'll fill in many of the on-camera scenes of the talking head with their own "perfect" images of what's being said.

Charisma, or stage presence, is hard to define, and all of us tend to perceive it somewhat differently. To one person, a talking head may exude charisma. To another, the same talking head is a dullard. It's a matter of how we relate—how we feel about the talking head, topic, and surroundings. Some of the factors that contribute to charisma are:

- projection of authority
- sincerity
- understanding
- trustworthiness
- demeanor, mannerisms, élan, looks, and dress

The charisma of the talking head is critical in the talking-head scenario. The late Reverend Billy Graham, a charismatic speaker, was able to mesmerize television and live audiences in Germany, Scandinavia, and Africa, audiences who often understood English less well than they understood their home languages. The words his interpreters spoke were almost anticlimactic.

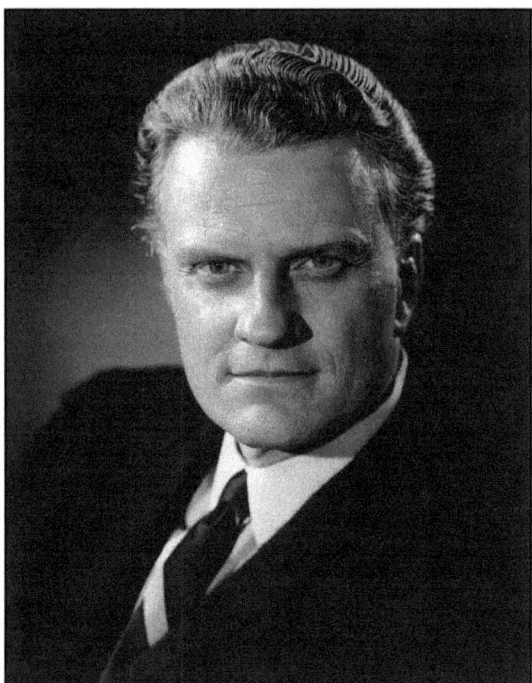

Figure 25. Reverend Billy Graham, a crusading evangelist.
Courtesy Larry Edmonds, Inc.

Clearly, Graham communicated to a host of different audiences. How did this happen? In communication theory, it is axiomatic that we tend to view, read, and talk about subjects that we already believe in, or want to believe in. So, additional communication on these subjects is reinforcement of what we already believe. Graham was a charismatic and inspirational speaker, and his messages were critically important to many in his audience. How about an atheist or agnostic? Would Graham have communicated to them? Perhaps. To answer the question, we need to know why such persons would want to watch and hear him. Some reasons might be curiosity, mental challenge, or peer pressure, or perhaps a need for the information or for inspiration. We'd need to know their preconceived notions about Graham and his message—hostile, neutral, or sympathetic. All these factors of the audience's mindset can and will influence the level of communication achieved.

In another example, some years ago the late actor James Whitmore portrayed President Harry S. Truman in a monologue performance in a simple setting. Audiences raved about the performance. They identified closely with and learned more about Truman than most would ever want to know. But we might ask, "Would an ardent fan of General Douglas MacArthur identify with and sympathize with Whitmore's interpretation of the Truman-MacArthur conflict?" I wonder.

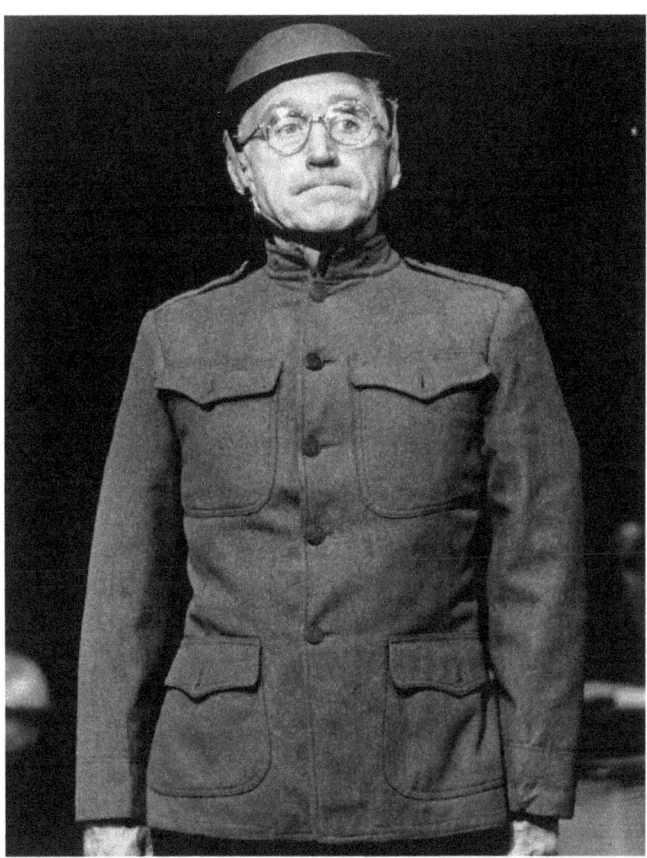

Figure 26. James Whitmore as President Harry S. Truman.
Courtesy Hollywood Books.

Exceptions

There are a few exceptions to consider. If the talking head is omnipotent or is the ultimate authority, and if the message transmitted is critically important to the audience, no matter how churlish or inept the talking head or how technically incompetent the show, a high level of communication most likely will occur because the audience will force themselves to pay close attention. They need the information for their advantage, and action items are required.

Here is one classic example of such an exception, long before we encoded information into digital formats and before the proliferation of social media. In 1980 the US Navy produced a ten-minute film to implement a tough, new, no-tolerance crackdown on drug abuse. A series of fatal aircraft accidents caused by personnel "high" on drugs precipitated this policy. The target audience was all naval personnel—admiral to seaman recruit. The show featured Admiral Thomas Hayward, Chief of Naval Operations, in a talking-head solo performance. The setting was his office. Production techniques were basic—one continuous ten-minute take from a fixed-camera position. No cutaways or visual aids of any kind were used. Occasionally, the cinematographer used the zoom lens. All the audience saw was the admiral talking directly to them—sailor to sailor—eye to eye—for the entire ten minutes. And the admiral was not a polished speaker; he stumbled several times. Ordinarily, such a film would communicate nothing to anybody. But this film was the exception. It was eminently successful. Why?

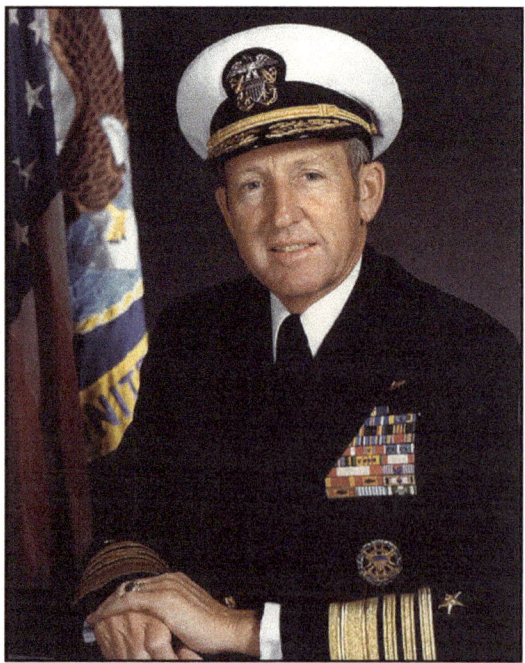

Figure 27. Admiral Thomas Hayward, Chief of Naval Operations.
Photograph courtesy US Navy.

Admiral Hayward cut an impressive figure in his service dress-blue uniform with its array of gold stripes and row upon row of decorations. He was the power—the father figure, the boss. His message was cogent, timely, and critically important. His sincerity, commitment, and concern projected right through the screen. He cared about each member of the audience, his shipmates, and the Navy. Empathy was intense. He made the severe consequences of any future drug abuse crystal clear. The audience got the message, and drug abuse in the Navy was reduced meaningfully.

Could this show have been made "better" with enhanced filmic design? Sure. Would such a "better" show have been more effective? Maybe. Would it have been as timely? No! There was an immediate need to distribute the message to the Fleet, now!

For the communication problem at hand, Admiral Hayward talked directly to the camera—to each audience member. This is a classic example of the message transcending all filmic-design inadequacies and the "unprofessional" delivery of the talking head. Perhaps the admiral's amateurism in front of the camera enhanced his sincerity and commitment to the naval audience. A classic case of "the message is the message." Though a professional actor might have given a more polished delivery, the actor probably could not have convinced the audience nearly as well.

Professional Actor

The professional actor has an important place in the talking-head scenario. It's the actor who is probably the better all-around storyteller. We see the actor in makeup, costume, setting, and with props. It's natural for us to project— we see the "real" character instead of an actor playing the part. It's the willing suspension of disbelief. We believe in the reality of the scenario. Reality is further enhanced by our prior knowledge about the character or subject. We provide imagery from our memory, triggered by visual and aural cues.

For example, in 1954, in a memorable solo performance, the actor Hal Holbrook played Mark Twain on stage for the first time. Later, his performance was broadcast on television. It's uncanny how deeply audiences felt that Holbrook *was* Twain! We saw and heard Mark Twain talking to us, in a scenario aided by nothing more than makeup, several cigars, a white suit, and a simple set. The show was an outstanding success—2,200 performances over sixty-three years.

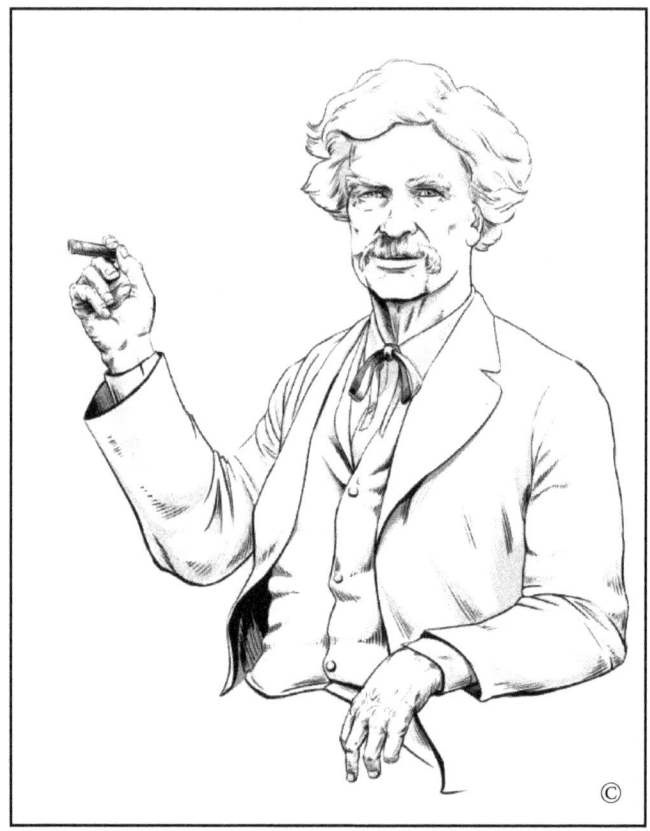

Figure 28. Hal Holbrook as Mark Twain. Doug Brown, commissioned artist.

Generally, I've found that it's best not to use actors as authority or power figures—unless, of course, they already are. Most actors in such roles are unconvincing, and audiences tend to resent the deception. In such instances, communication almost always fails.

Again, there are exceptions. In special situations, a professional actor is required. For example, when I produced the film *The Man From LOX,* I needed to use an actor to advance the narrative. The purpose of the film was to motivate liquid-oxygen (LOX) handlers to work safely. The selected setting was a fantasyland, nearly *avant-garde* in conception, in which an exception-ally gifted character actor portrayed the "Man" in a series of short, parodist vignettes. Only a skilled actor could have played this out-of-the-ordinary part with conviction.

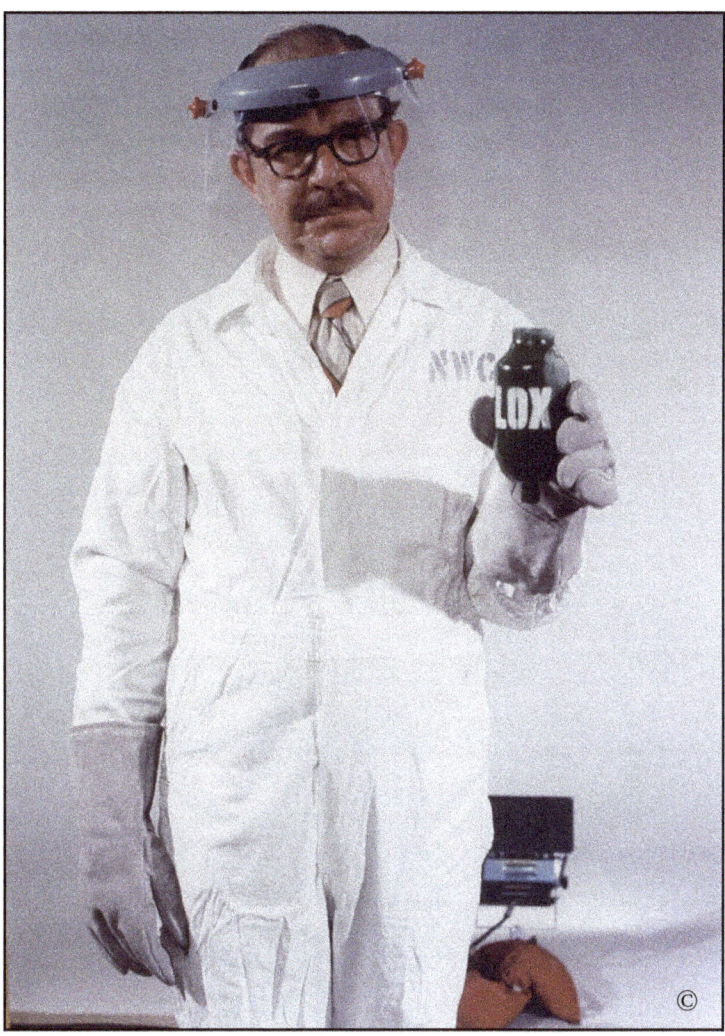

Figure 29. The Man From Lox (Lindsay Workman). Photograph from the author's private collection.

I heartily recommend that professional actors/narrators be used for off-cam-era narration—the anonymous voice of authority. No matter what the pressures are, don't compromise on this point. Audiences just won't accept an unprofessional narration, no matter who is doing it, and communication failure is guaranteed. Over the years, I have found that radio actors are excellent narrators.

Conclusion

The talking head *mise en scène* forfeits most of the communication power of our dynamic visual medium. Nonetheless, in some instances, I recommend this scenario. It's personal, quick, and economical, and it is an appropriate way to transmit critical messages quickly to concerned audiences.

Shelton's Pronunciamento
Use the talking head *mise en scène* with judicious skill and discretion.

Chapter 20

The Sound Track: an Integral Part of Information Motion-Media

Sound is a key dramatic element creating the ambience of our shows. Elements in the sound track are voice, music, and sound effects—although, with computer-controlled sound generators, it's difficult to distinguish between music and sound effects, and even sometimes between sound effects and voice.

Sound's power lies in its invisibility, its unobtrusiveness. Sound in our shows should be natural and inconspicuous. Our audience should not "hear" the sound. If our audience focuses on the sound, they will miss the messages transmitted by the kinetic visuals. For instance, wall-to-wall narration and wall-to-wall elevator music soon become noise in the communication process, no matter how expertly done. Ditto with overly done sound effects. An inept narrator or amateur players will generate lots of noise. It's to our advantage, therefore, to eschew amateur talent, off- and on-screen, and to choose professionals to narrate and act in our shows. We must resist to the depths of our professional being pressure from our client to put the CEO or whomever on camera to tell us all about "it." There are exceptions, as I've noted in the previous chapter.

Effective Sound Track

What's critical is that the sound and visuals work in harmony. If the audience notices that a sound doesn't "sound right," then the illusion of reality is broken and empathy is lost. It's critically important that music and sound effects not intrude on the commentary. If they do, they become noise that interferes with the show's messages. (In the lingo of our profession, we dubbed such a miscue a "hot mix.") Of course, if the sound elements are clear of commentary for a while, then "pot" them up to an appropriate volume. I've seen far too many shows in which the volume of music and sound effects intrudes into the commentary and destroys the show's effectiveness.

Shelton's Pronunciamento
Sound must be natural to the scene and not "heard."

To be unobtrusive, sound must mimic the perspective of the scene. For instance, if our scene is a long shot of an automobile driving on a lonely desert road, the volume, pitch, and tone of the automobile sound effect must seem as if it's emanating from the automobile and heard from the camera's (i.e., our) perspective. That is, the sound effect our audience hears ought to sound as if it comes from a microphone near the camera, not from a microphone on the automobile or on the side of the road as the automobile whizzes by.

Maintaining proper sound perspective is challenging. However, most sound houses have the wherewithal to modify sound effects to meet the scene's requirements. Naturally, all of this effort costs money. But I've found over the years that I kept postproduction expenses within budget by working with a professional sound house. It's truly amazing what wizardry these artists and engineers accomplish.

Silence can also be a telling dramatic effect. Such is particularly the case when a period of silence is broken by a startling and unexpected musical sting or an unforeseen sound effect—or the reverse, when music or effect quickly cuts to silence. The old adage "silence is golden" has much merit for our shows. Bela Balazs, filmmaker par excellence and author, notes that "no other art can reproduce silence."[1]

Types of Sound

Various sounds have an important role in setting the ambiance and pace of our shows. Here are some of the many factors involved:

- Voice talent is male, female, or obviously computer talk.
- Commentary intonation is hard or soft, fast or slow.
- Music is fortissimo or pianissimo or something in between, presto or adagio, euphonious or cacophonous, staccato or legato.
- Sound effects that are appropriate add realism and presence to a scene.

Note: if you use one sound effect in a scene, all elements in the scene must have appropriate sound effects. Otherwise, the scene calls attention to itself and your audience's suspension of disbelief will be broken.

A scene in which shots are edited in rhythm to a musical beat, or sound effects, is particularly effective in setting the scene This is especially true in a montage sequence. Or, we can do it the other way around: have music composed (or find "canned" music) that matches the tempo of our edited visuals.

Music

We can use a musical theme to bind disparate or widely separated scenes with continuity. Notice that we introduce music into a scene gradually. In the business, we dub this technique "fade-in." Ditto the reverse, a "fade-out." When we introduce a different music selection between scenes, we "segue" the tunes seamlessly: outgoing tune fades out while simultaneously the incoming tune fades in.

How do we obtain music to support our shows? The first thing to remember is that we must pay, one way or another, for the right to use any piece of music. It's best to follow the rules religiously.

We have three options to obtain music. But, first, please understand that I'm not an attorney. The laws on these matters are complex and vary from country to country. Always seek competent legal advice.

Original music. Commission a composer to write the tunes, contract with an arranger to orchestrate the music, hire a band or an orchestra to record the music, and book a sound studio for the recording session. The expenses involved with this task are considerable. If you choose this route, it's imperative that you get a "work for hire" release from everyone involved in the process. Persons who sign work-for-hire releases relinquish all claim to whatever work they have contributed to the project. In contributing their original work, they forgo their ownership of the copyright to it. Make no exceptions, or else you or your client may well be paying royalties every time your show is screened.

Copyrighted Music. Buy the license to use the music from the copyright owner, composer, publisher, performers, etc. This process is a tough, tedious, and expensive task fraught with legal proscriptions. There are, however, copyright organizations that specialize in making these kinds of deals. If the deal can be made, the cost may well be beyond our budget. Some of this copyrighted music is virtually unobtainable because of artist and publisher concerns or perhaps union restrictions.

It's _illegal_ for us to steal music and incorporate it into our motion-media show (from a CD or LP, for example.) Such a crime violates federal copyright and piracy laws, and would invite expensive lawsuits and fines that would denude our savings accounts and cause serious headaches. If the violation is egregious, time in the lockup may be appropriate.

Copyright-infringement laws apply to _any_ use of someone else's creative property. It matters not whether we use stolen music in a company-produced video to encourage employees to work safely, or in a TV commercial to publicize the Acme Company's new XYZ widget, or for any government or nonprofit institution's show, it is illegal, no matter who and no matter for what purpose. *No exceptions!*

Shelton's Pronunciamento
Never use music from commercial recordings without legal permission.

After a time, copyright protection expires and the musical work enters the public domain. In such a case, we're free to use the work *as originally published.* Accordingly, we need to hire an arranger, band or orchestra, sound house, etc. Congress sets this public-domain time limit and changes it from time to time. The copyright laws of other countries vary widely from those of the United States. Consult competent legal advice to ensure that the music you want is, in fact, in the public domain.

Music library. Purchasing music from a music library offers a number of advantages. It's easy, quick, and less expensive than the other options. Getting the licenses is routine. However, we need to make sure that the type of license we purchase covers the kind of show we've produced and the distribution method. (More on this theme in the "License" paragraph below). The amounts and kinds of music available in these libraries are enormous, and the performances are high quality—just about anything we'd ever want. Okay, it's not original music composed especially for our show, but it's relatively inexpensive, appropriate, and maybe even artistic! The type of music offered by these libraries is sometimes called "production music," since it is used for films, videos, multimedia, and other motion-media shows (the distribution scheme).

Another valuable advantage of music libraries is that the music is designed for motion-media use. The music is easier to edit because it has been scored with editing in mind—an important factor when we're trying to match an exact scene length timed to the tenth of a second.

We'll want to evaluate carefully several music libraries for the value they offer. Merely providing numerous themes is not enough for the discerning motion-media designer. The key is to find a library that offers the type of production music we want—including, perhaps, several edited or remixed alternate versions of a tune or theme, acoustic and electronic sounds, and sound effects. The size and depth of a library are critical; it should offer a wide variety of settings, periods, and moods.

Licenses. Depending on the music library, we may have a choice of several types of license. What we pay for is the license to use the music we select. We've have several options:

Annual or long-term license. Buy the license for the entire library, or a part of it, for a year or some other specific time and at an agreed-upon exhibition use. In such an instance, we take possession of the library in whatever medium it's been recorded. This option gives us a quantity of music with rights to use any of it in our shows. We log what tunes we've used and report our use to the library. Sometimes we may negotiate a deal in which there are no reporting requirements.

Individual license. Buy a license to use the music in a specific show and for a specific exhibition or distribution: internal use only, commercial television, worldwide sales meeting(s), rental or sales to users, etc. Sometimes there's a time limitation: no use beyond a set number of years without additional payments. We buy this kind of music license by the cut, the minute, the show, or the use, depending on the arrangement we make. In the ol' days, this type of agreement was known as a "needle drop." This term comes from the time when we placed a mechanical stylus on a vinyl long-playing record.

No matter which method we choose to get music, we must make sure that we have a 100-percent right to use the music in our show and in the exhibition manner we've agreed upon. To this end, we must have:

- Ownership, or
- Synchronization rights (to synchronize the music with visuals or spoken words and to reproduce copies); and
- Performance rights, to publicly perform the music in our show.

Sound Editor

A skillful music and sound effects editor is a treasured professional. They have knowledge of various music and sound effects libraries. They are instrumental in selecting the music and sound effects that blend optimally with our shows. I well remember, back in the analog days, working with music editors who could edit a tune to fit a scene: take an element from one tune and blend it into another, extend the time of a tune or contract it, and make the tune sound as if it were recorded that way. Today, with digital sound, such sorcery is routine.

Chapter 21

Distribution

Technology has radically changed the distribution of information. Today, we use an array of computers to distribute our show over the World Wide Web and any number of social media sites and electronic communication systems. In the future, who knows what systems we'll use?

Distribution scheme

Let's discuss several key factors in the distribution scheme, using these questions:

- What media do we recommend or does the client prefer?
- If the show is in a physical medium, how many copies are needed, and how do we get them to the audience?
- What arrangements need be made for social media and/or satellite distribution?
- Will the sponsoring organization distribute this show, or hire a professional distribution organization?

If the show is produced in CD/DVD form, it may be viewed by individuals on a personal computer, or played for groups in a theatre setting.

In this short chapter, I've highlighted just a few of the several permutations of the distribution scheme. Because technology is developing so rapidly, it's not prudent for me to delve into distribution any further. It's beyond the scope of this book (and my skill level).

For one example, nowadays, there may be no image or sound storage device in the motion-picture theatre. Synchronous satellites transmit digital images and sound to the theatre, office, classroom, home, most anywhere.

Last thought: I wholeheartedly recommend that the motion-media production organization defer from involvement in the distribution scheme—unless, of course, that's its forte.

Chapter 22

Our Client Isn't the Enemy

The information motion-media industry is big business and growing apace. Unfortunately, all too often, an adversarial relationship develops between the client and the script designer or producer. As a result, our show suffers. Sticking our egos in our pockets and following a few simple professional guidelines enhances understanding and makes our tasks of designing and producing information motion-media shows easier, lots more fun, effective, and profitable.

Essentially, our industry is not that much different from any other enterprise. A customer buys a product from a merchant at an agreed-upon price and delivery date. The customer is our client, the product is equipment, services, or a complete show. The show can be tailor-made or delivered "off the shelf." The merchant or retailer is the script designer or producer—you and me.

Selecting a Designer/Producer

Most motion-media shows are unique, custom-tailored to meet the client's singular needs. As a one-time event, a motion-media show is relatively expensive, has high visibility, and is not returnable—and there are no guarantees. How then do our clients select someone to design and produce their shows? It's not easy. The selection has to be right the first time. Selecting a designer/producer is much like selecting an architect to design an expensive home: it has to be right the first time.

Usually the first items a prudent client checks are our professional reputation, fiscal responsibility, and character—all details about us available on the World Wide Web for a few dollars. Here are key examples of what a potential client seeks to know:

- What recommendations do our previous clients make?
- How do we get along with clients?
- How do our peers regard us? Vendors?
- What is our work ethic?
- What experience do we have in relation to the proposed project?
- What sort of organization do we have—full-fledged or shoestring?
- How's our credit?
- Are we members of professional organizations? If so, which ones, and how involved are we?
- What have we contributed to our profession? Formal papers? Technical articles? Mentoring programs?
- The potential client is likely to ask to view and critique the shows we've produced or scripted.
- In checking with previous clients, they're likely to ask such questions as
  - Were communication objectives met to full measure?
  - Did we design/produce the shows with a judicious use of resources?
  - Were the shows delivered on time and within budget?
  - How readily were changes accomplished?
  - Did they enjoy working with us?
  - Would they hire us again?

Lastly, the prospective client must have an intuitive feeling that we're right for the assignment. The "chemistry" between the client and the producer/designer has to augur well for success before we reach an agreement.

Clarity Is Paramount

No matter how simple or complex the agreement—oral or written—we've established a contract. If we and our client conduct ourselves professionally, the goals of each are fulfilled. The completed motion-media show solves the client's needs, we've made a reasonable profit, learned something, had some fun, and enhanced our client base for repeat business.

Such a scenario seldom develops so favorably, however. Problems arise from a dearth of clear communication or common-sense business practices, or from unrealistic promises and expectations. Sometimes, problems occur from unethical behavior, the client's or ours. All too often, in the natural sequence of events, we can count on such problems to develop into disputes over:

- Budget and cost control
- Unplanned changes
- Missed or late delivery
- Creativity versus message
- Who said what to whom, when?

The ethics committee of the Producers' Guild notes that "usually the blame [for cost overruns, missed schedules, and poor results] is two-way. We as clients are not totally sure of what we want; [and] vendors will sell us whatever they think we will buy."[1]

It's imperative, therefore, that we make our contract "right" before beginning the project. When problems do arise, we've got to resolve them amicably and quickly according to the terms of the contract. Otherwise, we've begun an adversarial relationship with our client, now perceived as the enemy. We're now in a no-win predicament.

Unfortunately, in such hostile scenarios, the motion-media show's communication effectiveness is significantly reduced. And it's the communication business we're supposed to practice! Sometimes our egos cloud our understanding of just what we're supposed to be doing. We lose perspective on our common goal.

Divergent Goals

In this "techno/artistic" profession of ours, it's easy for the client's and producer's goals to diverge. I've frequently found that much of the divergence stems from the interpretation of "creativity" and "communication." The designer/

producer may feel that the client is stifling his/her artistic abilities, resulting in yet another "nuts-and-bolts" show. Conversely, the client may feel that the producer is gouging the budget and jeopardizing the schedule by creating another *Heaven's Gate* (one of the all-time, super-expensive movie-extravaganza flops), massaging his/her creative ego, and aiming at winning another award rather than focusing on communication.

As John Grierson said, "People are good enough at making films but not at using them to shape ideas, to dig out the heart of the matter. And in fact, the client may well be fearful and mistrustful of anything to do with art and aesthetics." Grierson often tried to avoid the words *art* and *documentary* in favor of *information* and *public service*.[2]

Cap Palmer noted, "The aim is the residual impression we want the film to leave—what our viewers will think or feel, perhaps subconsciously, as they leave the viewing experience and hopefully for some time afterward."[3]

Interpersonal Communication

Our success in the motion-media communication business is a function of our interpersonal communication skills—the ability to deal successfully with clients, crew, financiers, and others. Though our designing/producing skills are also critical, we'll not have the opportunity to use them unless we've communicated with prospective clients and convinced them that we can best solve their communication problem with our talent, expertise, integrity, and business acumen.

Communication with our clients must be clear and open. It must engender mutual trust and an explicit understanding of what's to be accomplished and how it's to be accomplished. Prospective clients want to work with friendly and sincere professionals. Technical competence may open the door, but interpersonal skills will close the deal.[4]

Essential to successful interpersonal communication skills is <u>listening</u>. We must listen carefully to our clients to understand their concerns so that we can articulate an appropriate solution for their communication problems. Up front, listening is a difficult task that requires significantly more concentration than does talking. Since we understand spoken concepts much more quickly than the time it takes to say them, our minds tend to wander, to develop a private plan, to debate, to hear less and less of what our client is saying.[5]

Also, we tend to hear with our own bias, which skews our understanding of what's being said. Oftentimes, we hear what we want to hear. Difficult as it is, we need to keep our bias in our pocket/purse and not let it become noise in the conversation. Listening is actually a form of persuasion. People like to be listened to and understood; it demonstrates that you are interested in them and their problems.

As skillful listeners, we give feedback, verbally and through body language. An utterance such as "I understand" or "I hear you" and a head nod from time to time lets the client know we're still tuned in. Ask questions, restate key points, make ancillary comments—convince the client that you're sincerely interested in his/her problem and know how to solve it in a way that optimizes communication, costs, and schedule.

Shelton's Pronunciamento
Clear and forthright communication is mandatory.

Team Activity

Listening is only part of our communication responsibility. Communication is a two-way interactive function—a team activity. Our clients must articulate their needs if we are to listen and understand. Regrettably, for many of our clients, the only experience they have had with motion-media communication is viewing narrative television and motion pictures. As we know, both of these media impart a distorted concept of information motion-media. Many clients are honestly ignorant. They don't know what's best or how to tell us what they want or need, or how to suggest what they want to accomplish. Here's where our listening and professional skills come to the fore.

Client Responsibilities

We need lots of help from our client. Our client has serious obligations and responsibilities to fulfill, just as we do. We'll highlight four key points here, but the contract should detail the full particulars regarding client and designer/producer responsibilities.

A well-drafted contract bodes well for a successful relationship. The contract is a legal promise between client and designer/producer, detailing each party's obligations and rights—who does what, when, and for how much. Our client's responsibilities lie in four primary areas:

- Consistency. The client stays locked into the agreed plan. Except for an extreme emergency, nothing changes (schedule, scope, etc.). Our client has the authority to get the job done the way it was planned. No vacillating.

- Support. The client performs tasks in a timely manner; clears the way for us in his/her organization, ensuring cooperation; and makes people, equipment, locations, etc., available on schedule.

- Deadlines. Client has a clear approval process within the sponsoring organization and ensures that approvals—script, first cut, payments, etc.—are executed promptly. The client knows that delays add to extra cost.

- Hands off. Client lets us do our job without interference (for example, "Let's make a 'movie'!" or "My niece is an aspiring actress"). Meddling is expensive, blows budgets, and reduces the show's effectiveness.[6]

Client/Producer Relations

Designing and producing a motion-media show is a lot easier, more profitable, and more fun when we've established a harmonious relationship with our client. In large measure, it's up to us to create a positive relationship. Our clients are in our game, with our rules—a game and rules with which they, most likely, are unfamiliar. The impetus should come from us to set a positive and understanding tone.

In such an atmosphere of constructive partnership, our show has a reasonable chance of emerging with audience and critical appeal.[7]

Should the client override our recommendations—which is the client's prerogative—and propose a solution that we're convinced will be undesirable or foolhardy, we're obligated to express our opposing opinion. We express our opinion politely, cogently, and constructively. However, if our client insists on proceeding with such a solution, notwithstanding our protest, we are ethically obligated to seek a compromise (seldom attainable in such situations), to comply wholeheartedly and demand a change to the contract, or to withdraw from the project.

If our client proposes something unethical or illegal we must retire from the project. It's a matter of integrity. Just how much is our reputation worth? That's the question. In the long term, no amount of largesse can redeem a tarnished reputation. And we must always strive for the long-term gain. Sometimes, from the beginning of our negotiations we know that a particular client relationship is untenable. That is the exact time to say *adios*.

I don't have a magic formula that would guarantee a positive and harmonious relationship with clients. It takes a lot of maturity on our part to set the positive tone and to maintain our professionalism, no matter how excellent or inferior our client may be, and to ensure that our client doesn't become our enemy. Perhaps the most irksome of all client/producer activities, the one that taxes our professionalism the most, is the approval process and the inevitable changes it involves.

Approvals and Changes

As much as we dislike the approval process, it is essential. Approvals are the primary means by which we ensure that client and designer/producer are communicating, that we're fulfilling the terms of the contract, and that the show meets expectations. Without approvals, we'll not have feedback, and we'll be well on the way to professional ruin. No matter how competent we think we are, we need feedback.[8] It's all too easy to detour from our primary goal. Sure enough, someone just might have a relevant idea or see something we've missed.

It's tough to see our aesthetic work critiqued by the bureaucracy: the killer committees where all members feel obligated to say something to impress the boss; legal departments that are constantly in the defensive mode; and nascent executives who fear and avoid decisions. We're in a strange profession when a novice can tell an expert what to do.[9] But that is the nature of our business. The novice is our client.

In the designing/producing process, there are several key stages at which the client should approve our work, and approve it in writing. Though all shows are different, the primary approval stages are at the completion of:

- Communication analysis and preproduction plan (includes budget and schedule details)
- Script treatment
- First-draft script and storyboard
- Additional script drafts
- Final script and storyboard
- Production plan and schedule (with contingencies)
- Interlock with edited work-print picture and narration/dialogue tracks only
- Interlock with final edited picture and master-dub soundtrack (with all sound elements)
- First released video (or other medium)

With multimedia, the approval process is much the same, though the terms and process may differ somewhat—for example, instructional design strategy, flowchart, and beta testing. It's important to remember that the more approvals there are, the longer the schedule and the more opportunities for the client to recommend (impose) changes, whether justified or not.

In my script-approval sessions, I usually don't hand the script to the client. Instead, I provide a copy of the storyboard. As I review the kinetic-visuals detailed on the storyboard, I read the appropriate narration or dialogue, if any. My reason for this gambit is that clients almost always "read" the commentary and forgo reviewing the scene descriptions. Consequently, they expect all the information to be in the commentary. They read the script as if it were a publication for the audience to peruse. It's difficult to convince the uninitiated otherwise.

Critical hazards to brace for in the approval process are the almost-assured presence of the review committee. They're "killers," mostly composed of inefficient, time-consuming ne'er-do-wells. Do all you can to avoid them. Try, with all your persuasive skill, to get the client to designate just <u>one person</u> as the approving authority. (Good luck!)

Usually, however, we'll be stuck, suffering the whims of review committees. Significantly, the composition of these committees fluctuates, with a continual flow of new members *ad infinitum* and *ad nauseam*. Most haven't a clue about what's happening or why. Be assured, nonetheless, that they do know they're supposed to make suggestions for "improvements"— and they do! These approval committee members have an unfailing tendency to demonstrate their alertness and keen acumen by spotting faults, errors, or misinterpretations real or imagined—all with much hullabaloo. They'll nit-pick to distraction with requests for frivolous changes. Some of these pundits may well recommend changes that vector the product away from the original plan.

Another major problem with review committees is that they try to satisfy everyone, which satisfies no one. It's "let's have everybody's input so there won't be any complaints." This does nothing for anybody. Also, review committees almost always focus on what's wrong rather than on what's right. Parry the review-committee comments with keen wit, consummate skill, resolute purpose, and the <u>contract</u>. Such is the genius of the approval process.

Shelton's Pronunciamento

Changes are valid if the contract says so.

One strategy I use to mitigate negative mindsets in approval sessions is to encourage the client to look only for the positive points on the first review—those points that have particular appeal and communicate well. On subsequent reviews, we look for and discuss the negatives. It's then that we decide which negative points are important enough to justify change. Some changes are super-costly, especially those surprises seen at interlock or first delivery.

Resolution of requests for changes is a critical and challenging task. We can usually resolve requested changes readily if none of us has developed a private plan and we're still working toward our common goal. Needed are lots of understanding, willingness to compromise, and diplomacy.

Normally, we, the designer/producer, are financially responsible for changes or redox caused by technical problems, inept execution, or delays we've caused. Those changes dealing with content, design, scope, concept, technical accuracy, and the like are the client's responsibility.[10] I'm assuming that these basics were agreed upon in the contract.

Financial responsibility for client-requested changes concerning aesthetics is usually not clear-cut. Ideally, we'll work with our client to make the determination on a case-by-case basis. In such instances, common sense and professionalism must prevail—there's no ready answer.

Naturally, we have to protect ourselves. For instance, after delivery of the second-draft script and storyboard that incorporates all agreed-upon changes and additions, some script designers charge an hourly rate to make additional changes in scope, direction, etc. Such charges cover costs of conferences, research, travel, and designing time. In other words, charge extra for any script or storyboard drafts beyond the second.

Here's the ultimate gambit that will keep the number of changes low and improve the show's ultimate effectiveness: tell the client how expensive such changes will be, raising the cost above the contract budget.

The Budget Belongs on the Screen/Monitor

Spend the client's money judiciously. Spend it on brains and craftsmanship. Avoid costly frills and high-tech stuff that doesn't enhance communication. It's simple enough to produce a self-important, super-colossal extravaganza: all it takes is money and chutzpah. In that case, we've put the production cart before the content horse.[11]

But it takes brains, craftsmanship, and commitment to design and produce a motion-media show that excels, one that is markedly superior in communicating our client's message to the target audience with an economy of resources. Such a product has an inherent dignity. Content is relevant and expressed with an honest simplicity and an underlying sense of purpose. Regardless of budget, craftsmanship is masterful. There are no excuses for awkward direction, inept cinematography/videography, unintelligible narration or dialogue, incoherent editing, unbalanced dubbing, or sloppy laboratory or transfer work.[12]

I must say that I've screened all too many motion-media shows that are inadequately crafted. They reek of amateurism—apathetically conceived, shoddily executed, and hastily completed. Such inadequate craftsmanship introduces noise in communication. Significantly, inadequate craftsmanship is not unique to our generation. In 1949, John Grierson lamented that "the camera work is less fresh and moving today, the cutting (editing) less dynamic, the sound less exploratory and inventive than they were 10 years ago, nor is it by accident that the writing, in general, is terrible and the habit of work less satisfactory to all concerned."[13] (John, what's changed?)

Is such inadequate craftsmanship the symptom of an inherited disease, passed on to generations of designers and producers? Have we fooled ourselves into believing that we're more skilled than we really are? Has the idiot "boob-tube" anesthetized our senses into wholesale mediocrity and passivity? Or is it that we just don't have pride in our work? The "whatever" attitude pervades our disinterest in our profession.

Make your motion-media shows sing with professionalism. Build a reputation on quality: *Integrity. Reliability. Commitment. Sensitivity. Communication skills.* Ensure that you and your clients work together harmoniously—professional associates pursuing a common goal.

Shelton's Pronunciamento
The simpler, the better.

FIN

(Appendices follow)

Appendices

Appendix 1

Evaluating Motion-Media Shows

One of the best ways to understand the essence of our motion-media profession and to increase our script designing and producing skills is to view and evaluate the shows of others, especially the shows of those who've earned critical acclaim. Screening is not enough, however. We must analyze and critique (evaluate) such shows with a critical eye. Tackle this evaluation task with industry, objectivity, and verve. You'll find the experience most rewarding as you gain keen insight into the fundamentals of our profession.

Evaluating the work of others is serious business! Their work deserves our wholehearted professional commitment. As you might expect, I've developed some rules and guidelines to follow. Rules, tedious as they may be, are necessary to:

- Establish professional standards
- Maintain consistency in evaluation and scoring
- Protect the integrity of the evaluations

Evaluation Rules/Guidelines

Attached is a copy of the Motion-Media Evaluation Sheet. Please become familiar with it before you begin viewing and evaluating.

1. Before and after a show is screened, you'll be told the show's primary audience and communication objectives. If you need more information, just ask. If this information is not available, make an educated guess from the data you have and from what you can discern after you've screened the show.

2. Rely on the total scope of your experience as the basis for your evaluation.

3. Evaluate the show from the perspective, experience, values, and attitudes of the *intended audience—*not your own. Admittedly, this projection is very difficult, yet it's the most important criterion in the evaluation process.

4. Evaluate only what you see and hear. All else is irrelevant! Don't be concerned with the budget, or what the show could have been, or should be, or with the good intentions or effort expended. Only the end results as seen and heard are pertinent.

5. Be concerned with communication values rather than gratuitous gimmicks and flashy folderol.

6. Immediately after you've screened a show, read each of the ten evaluation criteria on the Motion-Media Evaluation Sheet and mark each criterion with a score from 0 to 10. Decimals are okay. A mark of 5.0 is average. Don't be timid about using any number within the total range.

7. It's critically important that you mark a score in each of the ten evaluation criteria. Not doing so results in an unfairly low evaluation. For instance, from time to time a show may have no or minimal artwork/animation, and therefore you can't make an objective evaluation of the "art/animation" criterion. In such a case, use the average score of the other nine evaluation criteria for the art/animation score.

8. **Fifty is the average score.** That is, 50 is the score a competent, professionally produced show should get. It's rare for a show to score above 90 or below 10—perhaps one or two a session. Infrequently, shows score in the 80s or in the teens.

9. The average score of all the shows you evaluate ought to be about 50. The 50 average can vary somewhat when the shows are extraordinarily proficient or extremely inept.

10. If the average of all your scores is much above or below 50, reevaluate your standards and seriously consider rescoring the shows. Feel completely free to change your marks on a show that you believe you've scored too high or too low.

11. I recommend that you use pencil. It's easier to change.

12. I encourage you to make written comments on the Motion-Media Evaluation Sheet. Then compare your scores with the narrative. Are they consistent? Inconsistent?

13. Evaluating a show is a learning experience for all of us. To this end, discussion is encouraged after a screening and during the scoring. Try to keep comments short and to a minimum during the screening.

Motion-Media Evaluation Sheet ©

Title_____ Evaluation Total_____

Target Audience_____

Purpose: To_____

Producer_____

Client/Sponsor_____

Evaluation Criteria

0.0 is total imperfection. 5.0 is average. 10.0 is perfection. Evaluate each of the ten criteria with a number from 0 to 10. Decimals are okay.

Subjective Evaluation

How did you like this show? Use a number from 0 to 10. Vent your personal prejudices, pro or con, here. Then use the remainder of this evaluation sheet to score the show in a detached, professional manner.

   Score _____

Communication Value

Consider the target audience for this show.

Objectives. How well does this show accomplish its stated objectives? Consider the appropriateness of this show as a communication tool for achieving the stated goals: goals that must be fitting, realistic, and worthwhile.

   Score _____

Theme. Does this show have a clear vision and tone that communicate a sharply focused central theme to the target audience? Are resources (capital, energy, and screen time) used optimally? Are the essential elements of information developed logically, clearly, succinctly, and in the proper tone? Is communication used to maximum effectiveness with minimum expenditure of resources?

   Score _____

Information. Is the bulk of the information in the visuals? Do the visuals enhance communication to full measure? Is the show's message communicated effectively through relevant <u>filmic design</u> and <u>visual continuity</u>? Do visuals and narration cohere?

Score_____

Approach/Empathy. Does this show have a fresh and imaginative approach that facilitates information flow? Is the approach <u>pertinent</u>? Is "creativity" basic and <u>deliberately concealed</u>? Will the target audience accept this show? Does the approach generate empathy to gain and hold the audience's attention and involvement? Does the approach reinforce audience interest and commitment?

Score_____

Audio. Do audio elements reinforce the visuals? Is narration or dialogue used only to tell the audience what they cannot perceive from the visuals yet must know for a complete understanding? Do music and sound effects contribute to communication?

Score_____

Technical Quality

Cinematography/Videography. Is the cinematography/videography technically excellent? Is the composition aesthetically pleasing? Does the photography facilitate communication by highlighting important information? Does the lighting set an appropriate mood? Does the lighting have continuity throughout individual scenes?

Score_____

Sound. Is the sound crisp and clear? Are words pronounced with proper emphasis and spoken at the appropriate pace and style? Do music and sound effects enhance the ambience of the show? Are they unobtrusive? Are all sound elements blended to achieve a harmonious show?

Score_____

Editing. Are visuals sequenced for optimum communication? Is the plasticity of the medium (manipulation of time and space) used optimally to maintain orientation and to facilitate appropriate information flow? Is the pace appropriately varied and does it reinforce communication objectives? Is screen direction used effectively to create harmony and dissonance?

     Score_____

Art. Is the art style appropriate for the tenor of the show? Are art elements and spatial relationships used to accentuate key points? Is perspective true? Are shading and highlighting used to create depth and for emphasis? Are form, mass, and color arranged in compositions that heighten communication?

     Score_____

Add your individual scores and put the total on the top right-hand side of the front page. Please make narrative comments in the space below to support your numerical evaluation.

Evaluator_____

Date_____

© Copyright Shelton Communications. **All rights reserved.**

Appendix 2

101 Classic Documentary Films

The films I've listed are <u>old</u>. Nonetheless, they are perennial classics and ex-emplify the principles I've detailed in this book.

Film data are presented in the following sequence:
- Film title
- Filmmaker(s)
- Length
- Release date
- Producing organization/sponsor (as applicable)
- Country of origin
- Significant award(s)

Aero Engine. Arthur Elton; 55 min., 1933, Empire Marketing Board, England.

American Time Capsule. Chuck Braverman; 4 min., 1968, Braverman Productions, U.S.A.

Ballet Adagio. Norman McLaren; 10 min., 1971, National Film Board of Canada, Canada. Bronze Plaque, Columbus Film Festival.

Ballet Robotique. Bob Rogers; 8 min., 1982, Bob Rogers and Company, U.S.A. Academy Award nomination, 1983.

Ballon Rouge, Le (*The Red Balloon*). Albert Lamorisse; 34 min., 1956, Films Montsouris, France. Academy Award, 1957; *Palme d'Or,* Cannes Film Festival, 1957; Gold Medal, Grand Prix of the French Cinema, 1957.

Battle of San Pietro. Major John Huston, U.S. Army; 37 min., 1945, Signal Corps of U.S. Army, U.S.A. (Later documentation reveals that Huston staged the battle scenes.)

Battleship Potemkin, The. Sergei Eisenstein; 75 min., 1925, First Studio of Gosinko, Union of Soviet Socialist Republics.

BBC: The Voice of Britain. Stuart Legg and Alberto Cavalcanti; 56 min., 1935, General Post Office Film Unit for the British Broadcasting Corporation, England. *Médaille d'Honneur,* Brussels International Film Festival, 1935.

Bead Game. Ishu Patel; 6 min., 1977, National Film Board of Canada, Canada. (Figs. 13, 14)

Belle Époque, La (*Paris 1900*). Nicole Védres; 91 min., 1947, Pantheon-Productions Pierre Braumberger, France.

Berlin: Die Sinfonie der Grossstadt (*Berlin: Symphony of a Great City*). Walter Ruttman; 70 min., 1927, Fox-Europa, Germany.

Bolero, The. Allan Miller and William Fertik; 26 min., 1972, Allan Miller Productions, U.S.A. Academy Award, 1973.

Bridge, The. Joris Ivens; 12 min., 1928, Capi-Amsterdam, Holland.

Chelovek S Kinoapparatom (*Man with the Movie-Camera, The*). Dziga-Vertov; 69 min., 1929, Vufku, Ukraine, Union of Soviet Socialist Republics.

Civil War, The. Ken Burns; 7 hours, 1989, Florentine Films and WETA (Public Television), Washington, D.C., U.S.A.

City, The. Willard Van Dyke and Ralph Steiner; 44 min., 1939, American Documentary Films for the American Institute of Planners, U.S.A.

Claymation. Will Vinton and Susan Shadburne; 18 min., 1978, Will Vinton Productions, U.S.A. First Place, San Francisco International Film Festival.

Coal Face. Stuart Legg and Alberto Cavalcanti; 10 min., 1935, Empo (General Post Office Film Unit), England. Médaille d'Honneur, Brussels International Film Festival, 1935.

Contact. Paul Rotha; 42 min., 1933, British Instructional Films for Imperial Airways, Ltd., Shell-Mex, and British Petroleum, Ltd., England.

Crac! Frédéric Back; 15 min., 1981, *Société Radio-Canada,* Canada. Academy Award, 1982.

Cummington Story, The. Irving Lerner and Helen Grayson; 20 min., 1945, U.S. Office of War Information, U.S.A.

December 7th. Commander John Ford, U.S. Navy; 34 min., 1943, Coordinator of Information (Office of Strategic Services) for U.S. Navy, U.S.A. Academy Award, 1943. (Fig. 15)

Desert Victory. Ray Boulting and Major David MacDonald; 60 min., 1943, British Army Film and Photographic Unit and the Royal Air Force Film Production Unit for Ministry of Information, England. Academy Award, 1943.

Diary for Timothy. Humphrey Jennings and Basil Wright; 39 min., 1945, Crown Film Unit, Ministry of Information, England.

Drifters. John Grierson; 50 min., 1929, Empire Marketing Board Film Unit, England.

Eagle Has Landed: The Flight of Apollo 11. 29 min., 1969, National Aeronautics and Space Administration, U.S.A. Ionosphere Award, Atlanta International Film Festival.

Easter Island (original title, *L'Ile de Pâques*). Henri Strock and John Ferno; 25 min., 1935, France/Belgium coproduction.

Enough to Eat. Edgar Anstey; 22 min., 1936, Associated Realist Film Producers for Gas, Light and Coke Company, England.

Entr'acte (*Between the Acts*). René Clair; 14 min., 1924, independent production, France.

Face of Britain. Paul Rotha; 19 min., 1935, Gaumont-British Instructional, Ltd., England. *Médaille d'Honneur,* Brussels International Film Festival, 1936.

Face of Lincoln. Wilber T. Blume and Dick Harbor; 22 min., 1954, Cinema Department, University of Southern California, U.S.A. Academy Award, 1955.

Farrebique. Georges Rouquier and Étienne Lallier; 85 min., 1947, L'Ecran Français and *Les Films Étienne Lallier,* France.

Fiddle-de-dee. Norman McLaren; 3 min., 1947, National Film Board of Canada, Canada. First Prize, Salerno Film Festival.

Fighting Lady, The. Louis de Rochemont; 61 min., 1944, Twentieth-Century Fox for U.S. Navy, U.S.A. Academy Award, 1945.

Fires Were Started. Humphrey Jennings; 74 min., 1943, Crown Film Unit of the Central Office of Information, England.

Flight of the Gossamer Condor. Ben Shedd; 27 min., 1977, Shedd Productions, U.S.A. Academy Award, 1978.

General Line, The (The Old and New). Sergei M. Eisenstein and Grigori Alexandrov; 90 min., 1929, Sovkino-Moscow, Union of Soviet Socialist Republics.

Granton Trawler. John Grierson; 11 min., 1934, Empire Marketing Board, England.

Harlan County, USA. Barbara Kopple; 103 min., 1975, Cabin Creek Films, U.S.A. Academy Award, 1976.

Housing Problems. Arthur Elton and Edgar Anstey; 17 min., 1935, Associated Realist Film Producers for the British Commercial Gas Association, England.

If You Love This Planet. Terri Nash; 26 min., 1981, National Film Board of Canada, Canada. Academy Award, 1982. (Fig. 16)

Industrial Britain. John Grierson, Robert Flaherty, and Edgar Anstey; 21 min., 1933, Empire Marketing Board, released by Gaumont British Picture Corporation, England.

John F. Kennedy: Years of Lightning, Days of Drums. Produced by George Stevens Jr., written and directed by Bruce Herschensohn; 88 min., 1964, United States Information Agency for the John F. Kennedy Center for the Performing Arts, U.S.A. Golden Eagle, Council of International Nontheatrical Events, 1965.

Kon Tiki. Thor Heyerdahl; 70 min., 1947, RKO Radio/Janus, Sweden.

Kornet er i Fare (The Corn Is in Danger). Hagen Hasselbalch; 9 min., 1944, *Nordisk Films Kompagni*, Denmark.

L'Amitié Noire (Black Friendship). Jean Cocteau; 20 min., 1944, independent production, France.

Land, The. Robert Flaherty; 44 min., 1941, U.S. Film Service for Department of Agriculture, U.S.A.

Las Hurdes or *Tierra sin Pan, (Land Without Bread)*. Luis Buñuel and Pierre Unik; 27 min., 1933, Spain.

Lefty. James Thompson; 55 min., 1980, DBA Entertainment, U.S.A.

Les Mystères du Chateau de Dé, The Mysteries of the Chateau of Dé. Man Ray; 26 min., 1929, independent production, France.

Listen to Britain. Humphrey Jennings and Ian Dalrymple; 18 min., 1942, Crown Film Unit of the Central Office of Information, England. Academy Award nomination, 1943.

Louisiana Story. Robert Flaherty; 45 min., 1948, sponsored by Standard Oil Company of New Jersey, U.S.A. (Fig. 17)

Magic Rolling Board. Greg MacGillivray and Jim Freeman; 14 min., 1980, MacGillivray-Freeman Productions, U.S.A. Gold Cindy, Information Film Producers of America, 1981.

Man of Aran. Robert Flaherty; 62 min., 1934, Gaumont-British, England.

Memphis Belle, The: A Story of a Flying Fortress. Lt. Col. William Wyler, U.S. Army Air Force; 43 min., 1944, Eighth Air Force of the U.S. Army Air Force, U.S.A. Special Award, New York Film Critics Circle Awards, 1944. (Fig. 18)

Moana: A Romance of the Golden Age. Robert Flaherty; 85 min., 1926, Famous-Players-Lasky (Paramount), U.S.A.

Motion Painting I. Oskar Fischinger; 11 min., 1947, U.S.A.

My Father's Son. Gerald T. Rogers; 33 min., 1984, Gerald T. Rogers Productions, U.S.A. Best of Show, Cindy Competition, Information Film Producers of America, 1985.

Nanook of the North. Robert Flaherty; 54 min., 1922, Revillon Frères, U.S.A.

Native Land. Leo Hurwitz and Paul Strand; 88 min., 1942, Frontier Films, U.S.A.

Neighbours. Norman McLaren; 9 min., 1952, National Film Board of Canada, Canada. Academy Award, 1953.

Nieuwe Gronden (New Earth). Joris Ivens; 23 min., 1934, Capi-Amsterdam, Holland.

Night Mail. John Grierson, Harry Watt, and Basil Wright; 23 min., 1936, General Post Office Film Unit, England.

North Sea. Alberto Cavalcanti and Harry Watt; 26 min., 1938, General Post Office Film Unit, England.

Nuit et Brouillard (Night and Fog). Alain Resnais; 31 min., 1955, Cocinor Films, France.

NY, NY. Francis Thompson; 16 min., 1957, Francis Thompson Agency, U.S.A.

Olympia: Fest der Völker (Olympiad). Leni Riefenstahl; 3 1/2 hours, 1936, Olympia Film, Berlin, sponsored by *Nationalsozialistische Deutsche Arbeiterpartei* for the Reich Ministry of Public Enlightenment and Propaganda, Germany.

Pas de Deux. (*Dance for Two*). Norman McLaren; 13 min., 1967, National Film Board of Canada, Canada. Academy Award nomination, 1968. (Figs. 19, 20)

Plow That Broke the Plains, The. Pare Lorentz; 29 min., 1936, U.S. Resettlement Administration, U.S.A. (Fig. 21)

Power and the Land. Joris Ivens; 39 min., 1940, U.S. Film Service for the Rural Electrification Administration, U.S.A.

Powers of Ten: A Film Dealing with the Relative Size of Things in the Universe and the Effects of Adding Another Zero. Charles and Ray Eames; 9 min., 1978, Ray and Charles Eames Productions for IBM, U.S.A. Gold Medal, Miami International Film Festival, 1979.

Prelude to War (first of seven films in the *Why We Fight* series). Major Frank Capra, U.S. Army; 54 min., 1942, Special Services of the Orientation Branch of the War Department, U.S.A. Academy Award, 1943.

Quiet One, The. Sidney Meyers, Janice Loeb, and James Agee; 67 min., 1949, Film Documents, U.S.A.

Redes (The Wave). Paul Strand and Fred Zinnemann; 60 min., 1936, Secretariat of Education, Mexico.

Regen (Rain). Joris Ivens; 15 min., 1929 (sound added 1932), Capi-Amsterdam, Holland.

Retour, Le (The Return). Henri Cartier-Bresson and André Bac; 34 min., 1946, independent production, photography by U.S. Army Signal Corps and U.S. Army Air Corps cinematographers, France.

Rien que les Heures (Nothing but the Hours). Alberto Cavalcanti; 45 min., 1926, independent production, France.

River, The. Pare Lorentz; 32 min., 1937, U.S.D.A. Farm Security Administration for the U.S. Resettlement Administration, U.S.A. World Prize for Best Documentary, Venice Film Festival, 1938.

Roma, Città Aperta (Rome, Open City). Roberto Rossellini; 103 min., 1945, *Excelsia Film,* Italy. Grand Prize, Venice International Film Festival, 1946.

School in the Mail Box. Stanley Hawes; 18 min., 1946, Australian News and Information Bureau, Australia. Academy Award nomination, 1947.

Shipyard. Paul Rotha; 24 min., 1935, Gaumont-British Instructional for Orient Shipping Line, England.

Silent Witness, The. David W. Rolfe; 55 min., 1978, U.S.A. Flaherty Award, Best Documentary, British Film Academy, 1979.

Skuggor Over Snon (Shadows on the Snow). Arne Sucksdorff; 10 min., 1945, Svensk Filmindustri, Sweden.

Song of Ceylon. Basil Wright; 40 min., 1935, General Post Office Film Unit for the Ceylon Tea Propaganda Board, England. *Prix du Gouvernement,* Brussels International Film Festival, 1935.

Spanish Earth. Joris Ivens and Ernest Hemingway; 55 min., 1937, Contemporary Historians, Inc., U.S.A.

Survival Run. Robert Charlton and Joaquin Padro; 12 min., 1981, independent production, U.S.A. Grand Prix, Informfilm European Business and Industry Film Festival, 1982.

Swinging the Lambeth Walk. Lyn Lye; 4 min., 1940, Associated Realist Film Producers for Travel and Industrial Development Association, England.

Target for Tonight. Harry Watt and Ian Dalrymple; 50 min., 1941, Crown Film Unit, Ministry of Information, England. Honorary Award, Academy of Motion Picture Arts and Sciences, 1942.

Ten Days That Shook the World. Sergei Eisenstein; 105 min., 1928, Sovkino-Moscow, Union of Soviet Socialist Republics.

Transfer of Power: The History of the Toothed Wheel. Arthur Elton and Geoffrey Bell; 21 min., 1939, Shell Film Unit, England.

Triumph des Willen (*Triumph of the Will*). Leni Riefenstahl and Albert Speer; 110 min., 1936, Universum-Film, A.G., Berlin, sponsored by *Nationalsozialistische Deutsche Arbeiterpartei* for the Reich Ministry for Public Enlightenment and Propaganda, Germany. (Fig. 22)

True Glory, The. Garson Kanin and Carol Reed; 85 min., 1945, Film and Photographic Section of Supreme Headquarters Allied Expeditionary Force for U.S. Office of War Information and Ministry of Information of Great Britain, USA/England. Academy Award, 1946.

Universe. Les Novros; 26 min., 1976, Graphic Film Corporation for National Aeronautics and Space Administration, U.S.A. Academy Award nomination, 1977.

Un Chien Andalou (*An Andalusian Dog*). Luis Buñuel and Salvador Dali; 16 min., 1929, independent production, Spain.

Urban Spaces. Paul Winkler; 27 min., 1980, independent production, Australia.

Victory at Sea. Henry Solomon, music by Richard Rogers; 26 half-hour programs, 1952, National Broadcasting System, U.S.A. Emmy Award, 1953.

Walk in the Forest. Randall Hood; 28 min., 1978, independent production, Canada. Best of Show, Cindy Competition, Information Film Producers of America, 1980.

Waverly Steps. John Eldridge; 25 min., 1948, Crown Film Unit of the Central Office of Information, England.

Western Approaches. Pat Jackson; 80 min., 1944, Crown Film Unit of the Central Office of Information, England.

Why Man Creates. Saul Bass; 25 min., 1968, Saul Bass & Associates for Kaiser Aluminum and Chemical Corporation, U.S.A. Academy Award, 1969.

299 Foxtrot. S. Martin Shelton; 10 min., 1977, Naval Weapons Center for U.S. Navy, U.S.A. Gold Camera Award, U.S. Industrial Film Festival, 1978. (Figs. 23, 24)

Zem Spieva (*Earth Sings, The*). Karek Plicka; 67 min., 1932, independent production, Czechoslovakia.

Appendix 3

Split Page and Storyboard Format ©

The Scarf: The Perennial Fashion Statement

Script by Louise D. Burnham
and S. Martin Shelton

Storyboard by
Olivia Frances

Communication Analysis and Motion-Media Communication Plan ©

Today's Date: _____

Proposed Title:

The Scarf: The Perennial Fashion Statement

Reason to Produce This Show:

To introduce Acme Fashion, Ltd. new scarf line into our nationwide boutique chain

Target Audience:

Upscale female shoppers keen on fashion

Purpose:

To motivate the audience to purchase Acme Fashion's new line of scarves

Target Audience Profile

Identification factors

Demographic: Youthful (20s to 40s), female, single or married

Socioeconomic: Cross-section of female population leaning toward affluent

Psychological: Aggressive, stylish, and sophisticated

Motivation factors

Anticipation: Piqued through television, social media, and print advertising, and our trade-show "pitch."

Importance of goal achievement: Fashion-conscious need to be on cutting edge.

Urgency of communication: High; fashion vogue changes rapidly.

Information currency/obsolescence: Six to twelve months

Predisposition factors

> S̲ponsor̲: Sympathetic.
>
> C̲ommunicator̲: Neutral.
>
> I̲nformation̲: Interested. The show keeps audience *au courant*.
>
> M̲edia̲: Positive. This is the electronic media generation.

Secondary Audience:

Wholesale dealers

Essential Elements of Information

1. Scarves and their derivatives have been used throughout history.
2. The scarf is a versatile fashion accessory.
3. We offer a potpourri of stylish scarves: sizes, colors, patterns, fabrics.
4. The quality of Acme Fashion scarves is preeminent.

Technical Quality Needed:

Technically excellent, sophisticated, and polished.

Interaction:

Acme Fashion sales associates

Schedule

Research: 7 days

Treatment: 5 days

Storyboard script 1st draft: 5 days

Final storyboard script draft: 3 days

Production: 14 days

Postproduction: 6 days

Duplication: 3 days

Distribution: 1 week

Filmic Approach

Tenor

Dramatic, upbeat, stylish, fast-paced, exciting

Milieu

Various historical and contemporary locations featuring models wearing, as appropriate, our new line of scarves

Characteristics

Motion medium, still photographs, models/actors, original contemporary music

Form

Current high-end production motion-medium.

Communication Surround

Audience size at each screening:

In our boutiques, audience range from one to four.

At our trade-show booth, dozens of viewers continuously

Frequency of screening:

Boutiques: when sales associates deem appropriate

Trade-show booth: continuous throughout the day

Physical environment of the viewing site:

Point-of-sale display at the scarf counter in our boutiques

Elegant booth at trade shows

Projection/viewing equipment:

Acme Technologies, Ltd. table-top media viewer

Projectionist:

Acme Fashion sales associates

Power requirements:

Standard for the area

Backup Material and Equipment:

Elegant point-of-sale display, models and manikins showing Acme Fashion's line of scarves, sample scarves, and color brochure handouts

Controlling Factors

Due date

Early November: in time for spring fashion-trade shows

Serialized

No

Part of total communication package

Yes. Extensive advertising campaign in appropriate fashion media, sample scarves to boutique buyers, point-of-sale handouts

Changes or updates

Yes. Models' attire needs to reflect seasonal offerings.

Technical and political production considerations

Need coordination and permission to film on Fifth Avenue in NYC and in the various other locations noted in the script

Hazards and safety considerations

Blocking off Fifth Avenue and other concerns we'll determine at each location

Client Concerns

Image projected

High fashion; Acme Fashion's scarves provide *haute* variety in all wardrobes.

Company/organization policy

Acme Fashion scarves are the vanguard of fashion.

Legal aspects

Music and still photographs clearances; model/actor releases; "Work for Hire" documents from all production personnel, liability insurance, and completion bond

Political impact

None.

Proprietary information

All Acme Fashion's scarf designs are proprietary information. Do not release before the first November trade show.

Classified information

No government information

Budget

Script/Storyboard: $65,000

Production: $350,000

Distribution: $75,000

Medium Selection

Producing medium:

Current high-definition, three-dimensional, holographic production medium

Foreign language versions

None. There's no spoken narration/dialogue.

Primary distribution medium

Current high-definition, three-dimensional, holographic distribution media

Secondary distribution medium

Social media, World Wide Web

Key Personnel

Client/sponsor

<u>Organization</u>

Acme Fashion, Inc.

<u>Names of contacts</u>

Ms. Penelope Worthington-Smythe and Ms. Brigid O'Shaughnessy

<u>Job title</u>

Marketing Directors

<u>Cell number</u>

(212) xx6-5000

<u>Email address</u>

xxx@yyy.net

Address

ONE Penn Square, Pennsylvania Station, NYC 10011

Technical Advisor/Subject Matter Expert

Name

Ms. Amelia Ebberhart-Putnam

Job title

Coordinator of the Liaison

Department

Public Affairs Department

Cell number

(212) xx6-5001

Email

zzzz@yyyz.net

Address

TWO Penn Square, Pennsylvania Station, NYC 10011

Client/Sponsor Approval Authority

Name

Ms. Madeleine de' Sunderman-Wo

Job title

Chief Executive Officer

Organization

Corporate office

Cell number

(212) xx6-5000

Email

www@yyyw.net

Address

THREE Penn Square, Pennsylvania Station, NYC, 10011

Producer

Name

Ms. Amanda Lupercio-Pirogoff

<u>Job title</u>

Executive producer

<u>Organization</u>

Super Colossal Productions, LLC

<u>Cell number</u>

(202) xxx-9999

<u>Email</u>

aaaa@bbb.com

<u>Address</u>

1776 Avenue Z, SSW, Washington, D.C. 20408

Script Designer

<u>Name</u>

Ms. Miles Archer-Spade

<u>Job title</u>

Script Designer Extraordinaire

<u>Organization</u>

Dial-A-Script, Ltd.

<u>Cell number</u>

202: xxx 8888

<u>Email</u>

cccc@ddd.com

<u>Address</u>

1775 Avenue X, NNE, Washington, DC 20410

The Scarf: The Perennial Fashion Statement ©
(Split-page format)

FADE IN	FADE IN MUSIC
Main Title. *The Scarf: The Perennial Fashion Statement*	(Music is soft jazz piano accompanied with base, snare drums, and xylophone.)
FADE OUT	
FADE IN	SEGUE
Scene # 1 A, B, C, and D. Start this scene with a MEDIUM SHOT of old black-and-white photographs of Isadora Duncan dancing. She wears scarves. Build montage of other photographs showing her dancing with scarves wafting about. Vary with CLOSE SHOTS, MEDIUM SHOTS, and LONG SHOTS.	(Music by Franz Schubert, as used by Duncan)
DISSOLVE TO	SEGUE

Scene # 2 A, B, C, D, E, F, G, and (Lively music to set the pace for
H. Montage of historic "scarves the montage)
on parade"—people from differ-
ent cultures wearing scarves: for
example, veils and dance par-
aphernalia on Middle Eastern
dancer; sari on Indian woman;
babushka on old Russian woman; kerchief on Swiss milkmaid;
ascot tucked into overcoat of
Dickens-era dandy; turban and
scarf on pre-Civil War female
slave; Western desperado wearing kerchief over lower face;
wraparound scarves on biki-
ni-clad South Pacific islanders.
Vary with CLOSE SHOTS,
MEDIUM SHOTS, and LONG
SHOTS.

DISSOLVE TO   SEGUE

Scene # 3A. CLOSE-UP in (Contemporary romantic
SLOW MOTION of extremely mood-music)
long chiffon scarf rippling in the
breeze against a sunset sky.

PAN RIGHT to reveal scene # CONTINUE MUSIC
3B.

Scene # 3B. MEDIUM SHOT. (Music UP slightly)
The scarf is part of an elegant
woman's evening attire. She is
with a male companion. They
are leaning against a cruise
ship's rail.

DISSOLVE TO

SEGUE

Scene # 4. MEDIUM SHOT of the upscale shopping district of Fifth Avenue in New York. The scene is busy with shoppers. Most are walking left to right.

(Big-band orchestra plays appropriate "city" music. Add sound effects: beeps, honks, traffic, and other city sounds.)

PAN RIGHT to reveal scene # 5

SEGUE

Scene # 5. MEDIUM LONG SHOT the camera concentrates on a solitary woman walking briskly screen left (opposite most of the foot traffic). She is outfitted with a yellow wool dress accented with a bright multihued floral-print-on-black-background scarf worn as a sash and tied at the waist. She is elegant; carries a briefcase; obviously she is a successful businesswoman.

(Solo saxophone plays a contemporary jazz tune that is counterpoint to the music in scene # 4. FADE OUT gradually the sound effects. Music UP as the scene plays.)

FADE OUT MUSIC

CUT TO

FADE IN MUSIC

Scene # 6. TIGHT CLOSE SHOT of woman looking into a mirror and casually adjusting an earring that is situated below her contemporary and attractive scarf, worn as a headband.

(The music is upbeat, soft contemporary-rock played softly.)

CAMERA DOLLIES BACK slightly to reveal more of the scene. The setting tells us that the woman is an artist. We see paints and an easel in the mirrored background.

CUT TO

SEGUE and MUSIC UP

Scene # 7A, B and C. Build a filmic design MONTAGE of MEDIUM CLOSE AND CLOSE SHOTS cut in rhythm to the semi-rapid beat of the music. Several women in different professional environments: bank, hospital, computer laboratory, kindergarten. They sport multi-colored scarves draped variously from front to back, and side to side, as shawls, bibs, and cowls.

(Contemporary jazz with a semi-rapid beat)

CUT TO

CONTINUE MUSIC

Scene # 8. MEDIUM LONG SHOT of woman cinching a long scarf as a belt. She is in front of a freestanding mirror.

(The beat slows.)

CUT TO

SEGUE

Scene # 9A, B, and C. Build film-ic design montage of EXTREME CLOSE SHOTS and CLOSE SHOTS of women roping and knotting scarves on themselves as neckties, bowties, rosettes, and collars.

(Music is soft jazz piano accom-panied by base, snare drums, and xylophone—same music as in the main title scene.)

DISSOLVE TO

CONTINUE MUSIC

Scene # 10. LONG SHOT of nattily dressed woman in our upscale Fifth Avenue boutique. She is browsing a display of our new line of scarves.

(Beat increases slightly)

MATCH ACTION CUT

CONTINUE MUSIC

Scene # 11A. EXTREME CLOSE SHOT of the scarf display. Our brand, Acme Scarf, Inc., is dis-played tastefully.

(Volume UP slightly)

MATCH ACTION CUT

CONTINUE MUSIC

Scene # 11B. MEDIUM SHOT of the woman selecting a mul-ticolored scarf. We see that our brand is sharply focused.

(As we build to the climax, the music is UP gradually.)

MATCH ACTION CUT

CONTINUE MUSIC

Scene # 12A. MEDIUM LONG   (The music is UP slightly more.)
SHOT of the woman noticing
that the colors of the selected
scarf complement her outfit.

MATCH ACTION CUT   CONTINUE MUSIC

Scene # 12 B. MEDIUM LONG   (Continue building the music to a
SHOT of woman. She begins to   climactic volume.)
drape and tie the scarf this way
and that. With each variation
she sees harmony.

MATCH ACTION CUT   CONTINUE MUSIC

Scene # 12C. MEDIUM CLOSE   (Music is UP.)
SHOT (TWO SHOT) of wom-
an and sales associate. Woman smiles, unties the scarf, and
hands it over the counter to the
sales associate to make a purchase.

ZOOM IN   CONTINUE MUSIC

Scene # 12 D. TIGHT CLOSE   (Music is UP FULL and ends with
SHOT of our name and logo   a flourish.)
that are clearly visible on the
scarf.

FADE OUT   FADE OUT MUSIC

(The storyboard for this show follows.)

The Scarf Storyboard ©

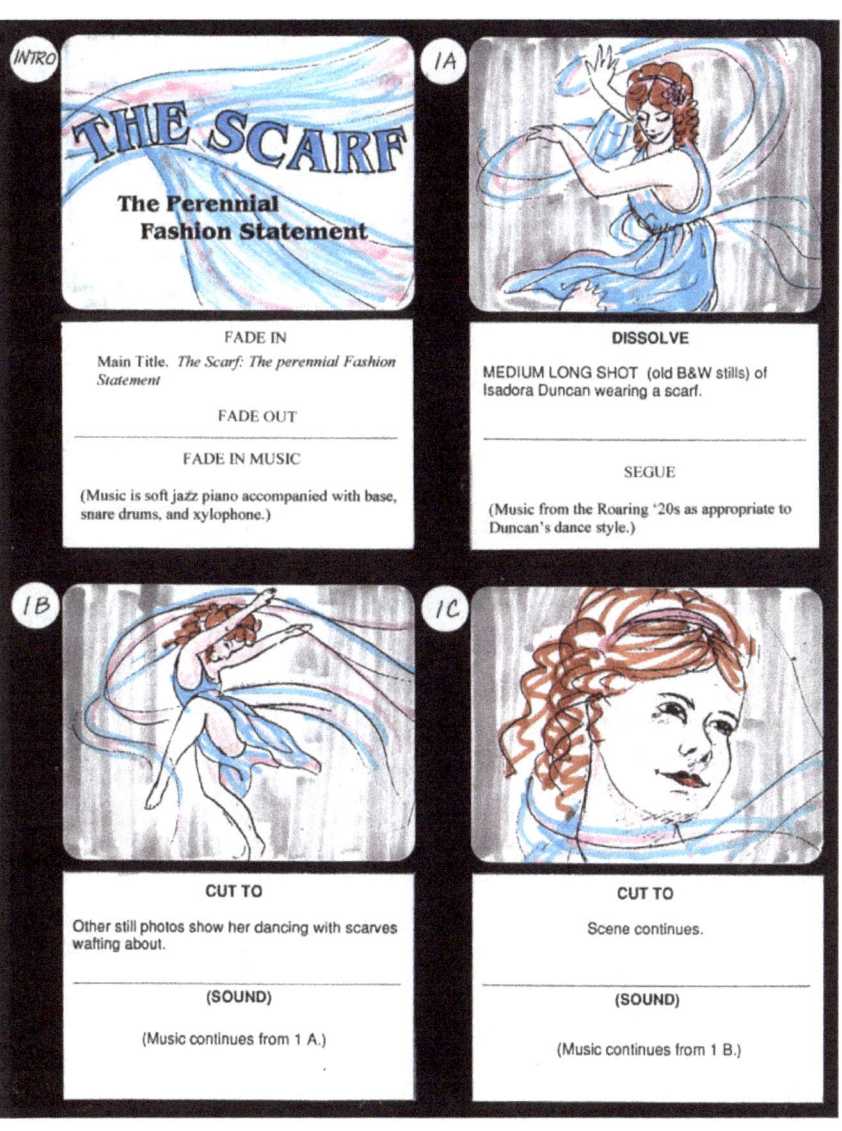

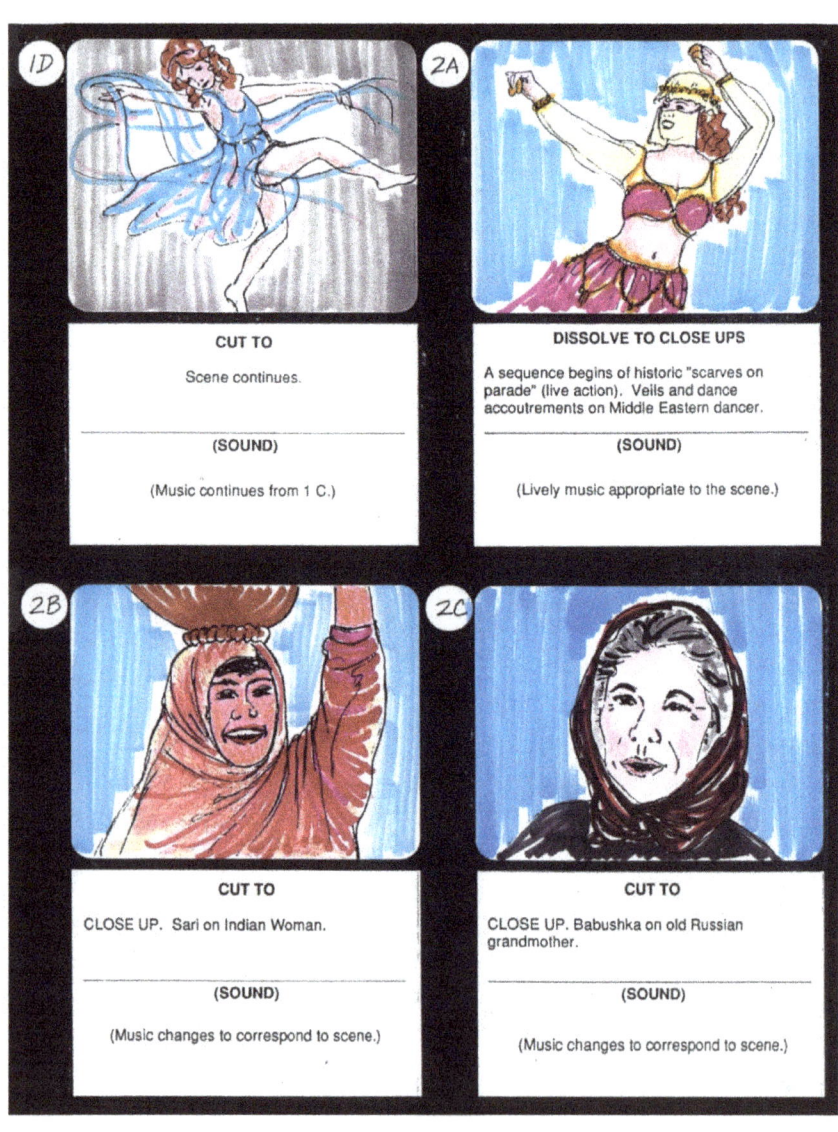

1D

CUT TO

Scene continues.

_____

(SOUND)

(Music continues from 1 C.)

2A

DISSOLVE TO CLOSE UPS

A sequence begins of historic "scarves on parade" (live action). Veils and dance accoutrements on Middle Eastern dancer.

_____

(SOUND)

(Lively music appropriate to the scene.)

2B

CUT TO

CLOSE UP. Sari on Indian Woman.

_____

(SOUND)

(Music changes to correspond to scene.)

2C

CUT TO

CLOSE UP. Babushka on old Russian grandmother.

_____

(SOUND)

(Music changes to correspond to scene.)

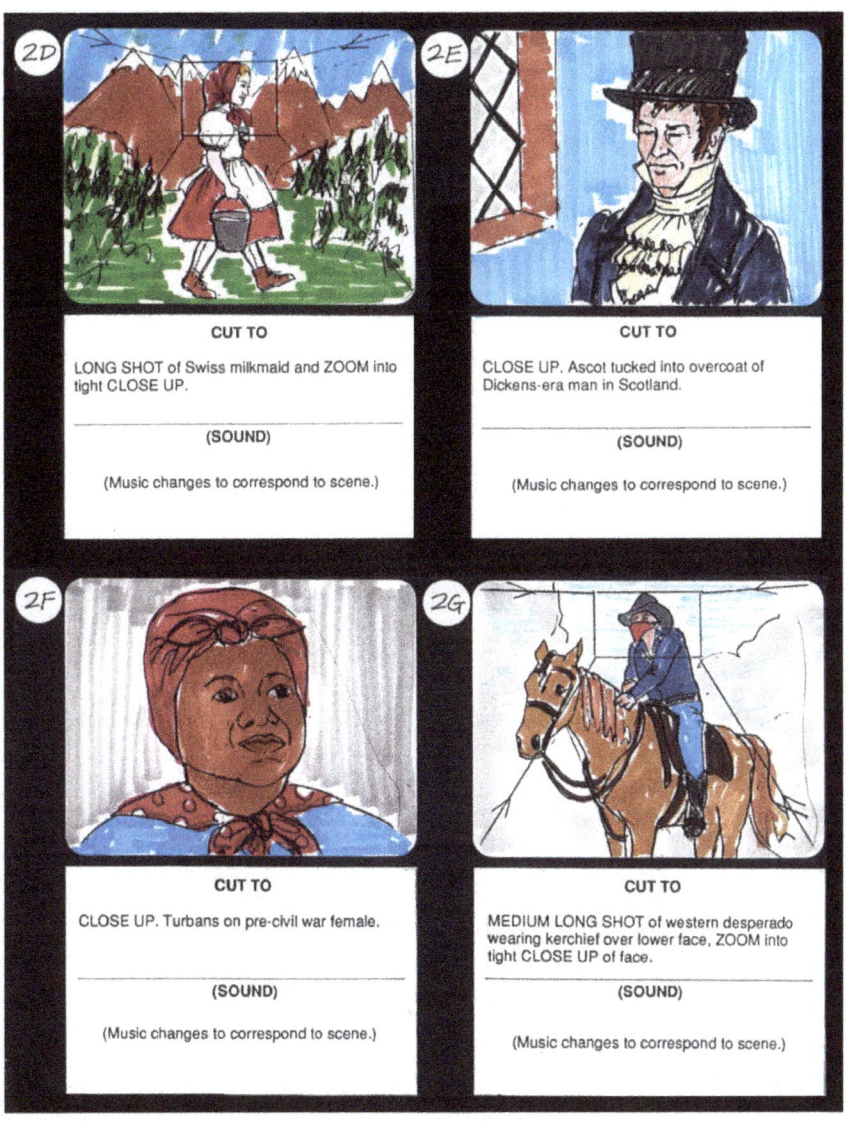

2D

CUT TO

LONG SHOT of Swiss milkmaid and ZOOM into tight CLOSE UP.

(SOUND)

(Music changes to correspond to scene.)

2E

CUT TO

CLOSE UP. Ascot tucked into overcoat of Dickens-era man in Scotland.

(SOUND)

(Music changes to correspond to scene.)

2F

CUT TO

CLOSE UP. Turbans on pre-civil war female.

(SOUND)

(Music changes to correspond to scene.)

2G

CUT TO

MEDIUM LONG SHOT of western desperado wearing kerchief over lower face, ZOOM into tight CLOSE UP of face.

(SOUND)

(Music changes to correspond to scene.)

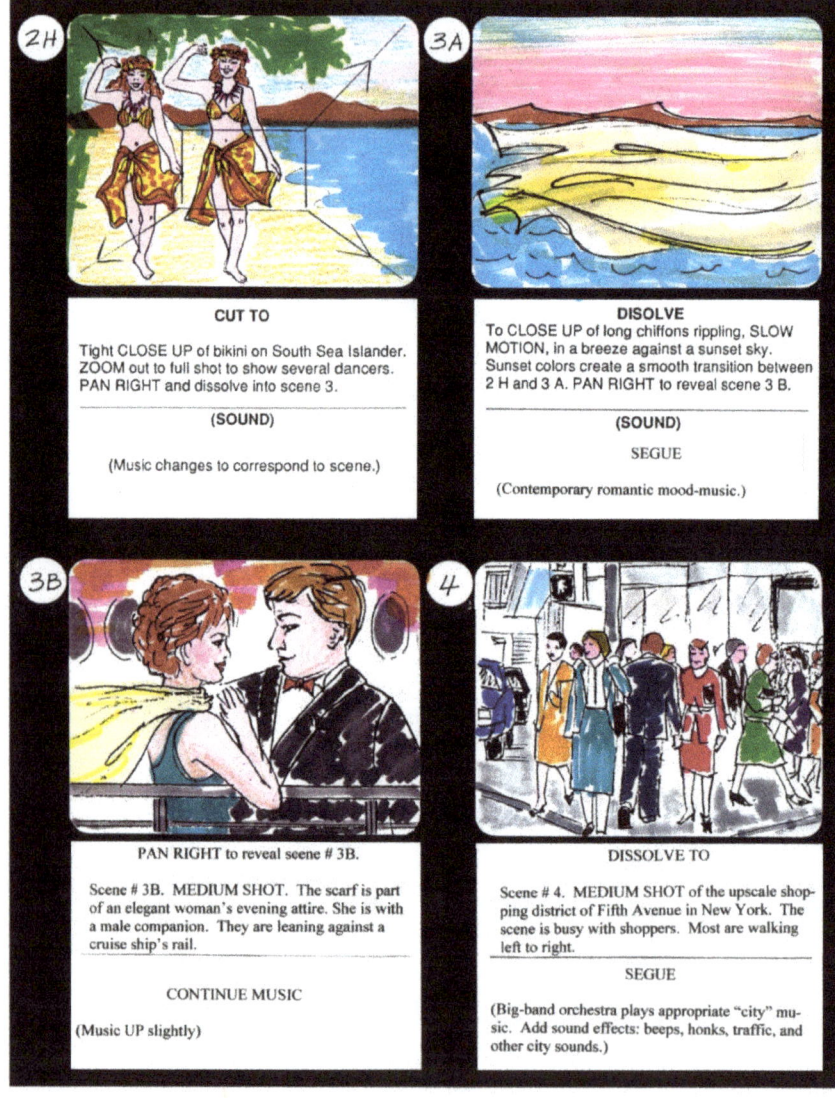

2H

CUT TO

Tight CLOSE UP of bikini on South Sea Islander.
ZOOM out to full shot to show several dancers.
PAN RIGHT and dissolve into scene 3.

(SOUND)

(Music changes to correspond to scene.)

3A

DISSOLVE

To CLOSE UP of long chiffons rippling, SLOW
MOTION, in a breeze against a sunset sky.
Sunset colors create a smooth transition between
2 H and 3 A. PAN RIGHT to reveal scene 3 B.

(SOUND)

SEGUE

(Contemporary romantic mood-music.)

3B

PAN RIGHT to reveal scene # 3B.

Scene # 3B. MEDIUM SHOT. The scarf is part
of an elegant woman's evening attire. She is with
a male companion. They are leaning against a
cruise ship's rail.

CONTINUE MUSIC

(Music UP slightly)

4

DISSOLVE TO

Scene # 4. MEDIUM SHOT of the upscale shopping district of Fifth Avenue in New York. The
scene is busy with shoppers. Most are walking
left to right.

SEGUE

(Big-band orchestra plays appropriate "city" music. Add sound effects: beeps, honks, traffic, and
other city sounds.)

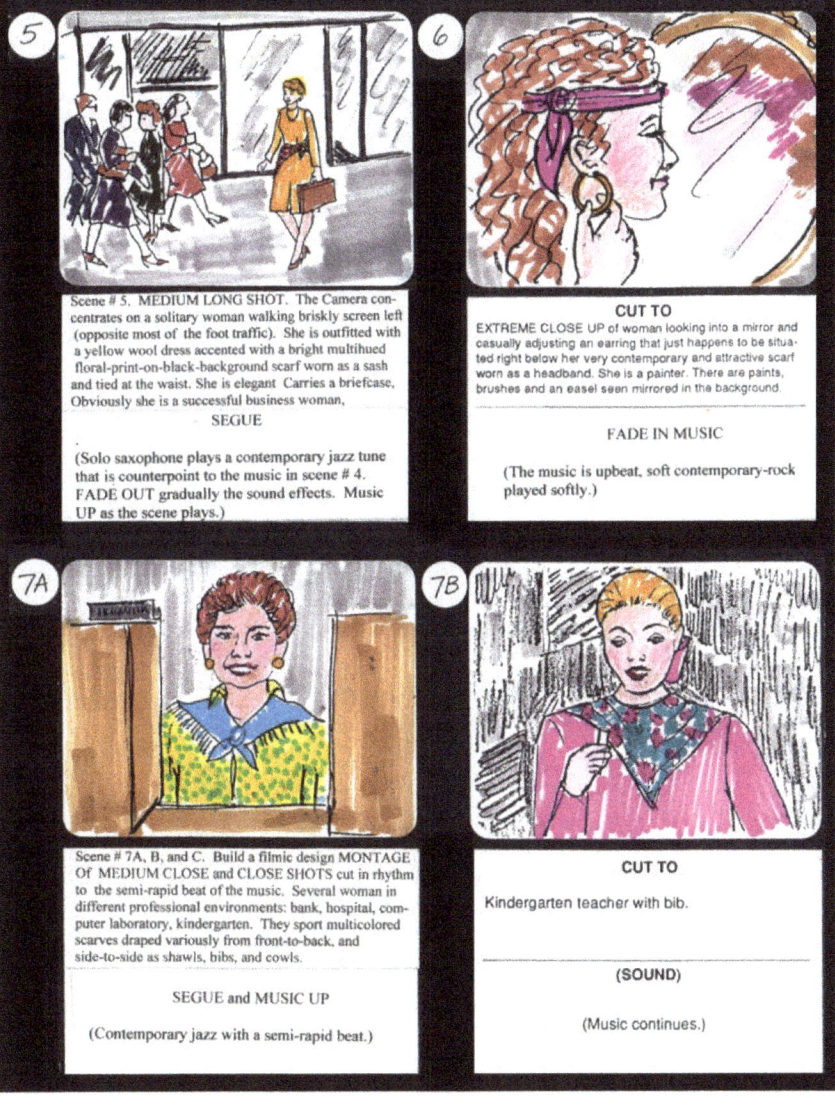

5

Scene # 5. MEDIUM LONG SHOT. The Camera concentrates on a solitary woman walking briskly screen left (opposite most of the foot traffic). She is outfitted with a yellow wool dress accented with a bright multihued floral-print-on-black-background scarf worn as a sash and tied at the waist. She is elegant Carries a briefcase, Obviously she is a successful business woman,

SEGUE

(Solo saxophone plays a contemporary jazz tune that is counterpoint to the music in scene # 4. FADE OUT gradually the sound effects. Music UP as the scene plays.)

6

CUT TO

EXTREME CLOSE UP of woman looking into a mirror and casually adjusting an earring that just happens to be situated right below her very contemporary and attractive scarf worn as a headband. She is a painter. There are paints, brushes and an easel seen mirrored in the background.

FADE IN MUSIC

(The music is upbeat, soft contemporary-rock played softly.)

7A

Scene # 7A, B, and C. Build a filmic design MONTAGE Of MEDIUM CLOSE and CLOSE SHOTS cut in rhythm to the semi-rapid beat of the music. Several woman in different professional environments: bank, hospital, computer laboratory, kindergarten. They sport multicolored scarves draped variously from front-to-back, and side-to-side as shawls, bibs, and cowls.

SEGUE and MUSIC UP

(Contemporary jazz with a semi-rapid beat.)

7B

CUT TO

Kindergarten teacher with bib.

(SOUND)

(Music continues.)

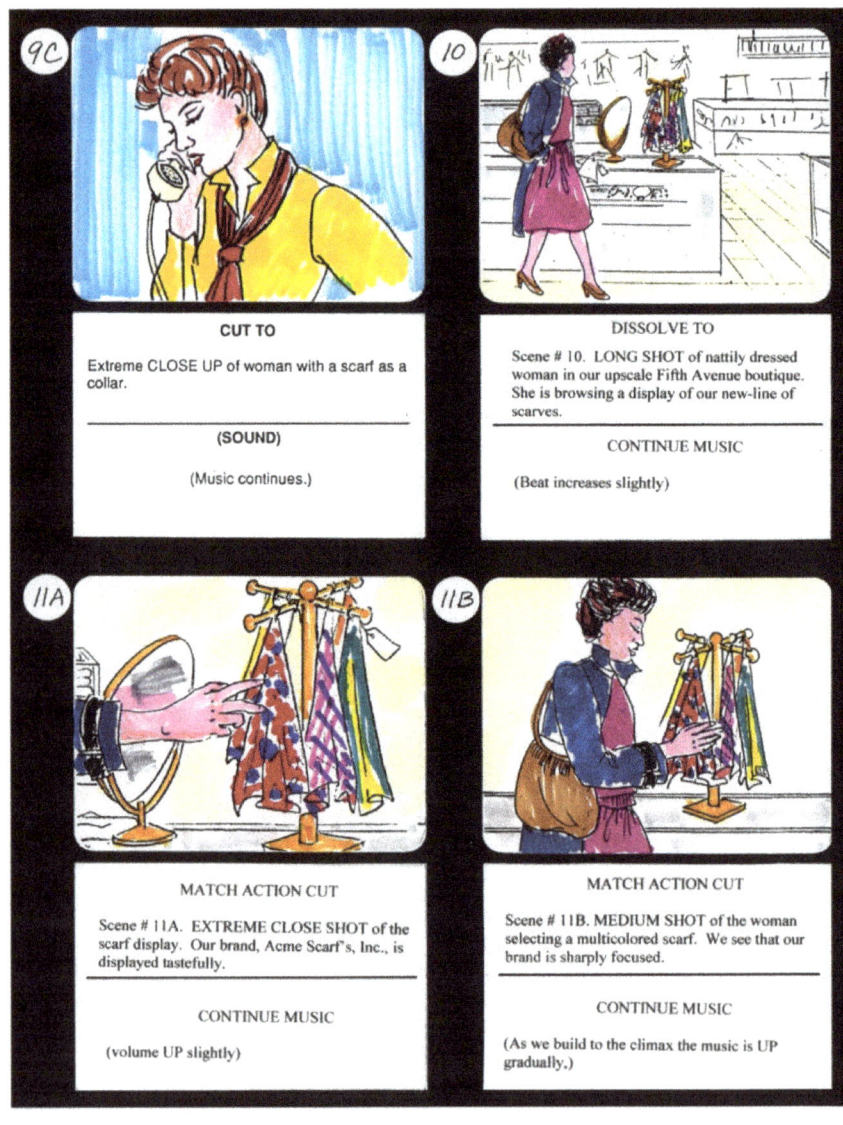

9C

CUT TO

Extreme CLOSE UP of woman with a scarf as a collar.

(SOUND)

(Music continues.)

10

DISSOLVE TO

Scene # 10. LONG SHOT of nattily dressed woman in our upscale Fifth Avenue boutique. She is browsing a display of our new-line of scarves.

CONTINUE MUSIC

(Beat increases slightly)

11A

MATCH ACTION CUT

Scene # 11A. EXTREME CLOSE SHOT of the scarf display. Our brand, Acme Scarf's, Inc., is displayed tastefully.

CONTINUE MUSIC

(volume UP slightly)

11B

MATCH ACTION CUT

Scene # 11B. MEDIUM SHOT of the woman selecting a multicolored scarf. We see that our brand is sharply focused.

CONTINUE MUSIC

(As we build to the climax the music is UP gradually.)

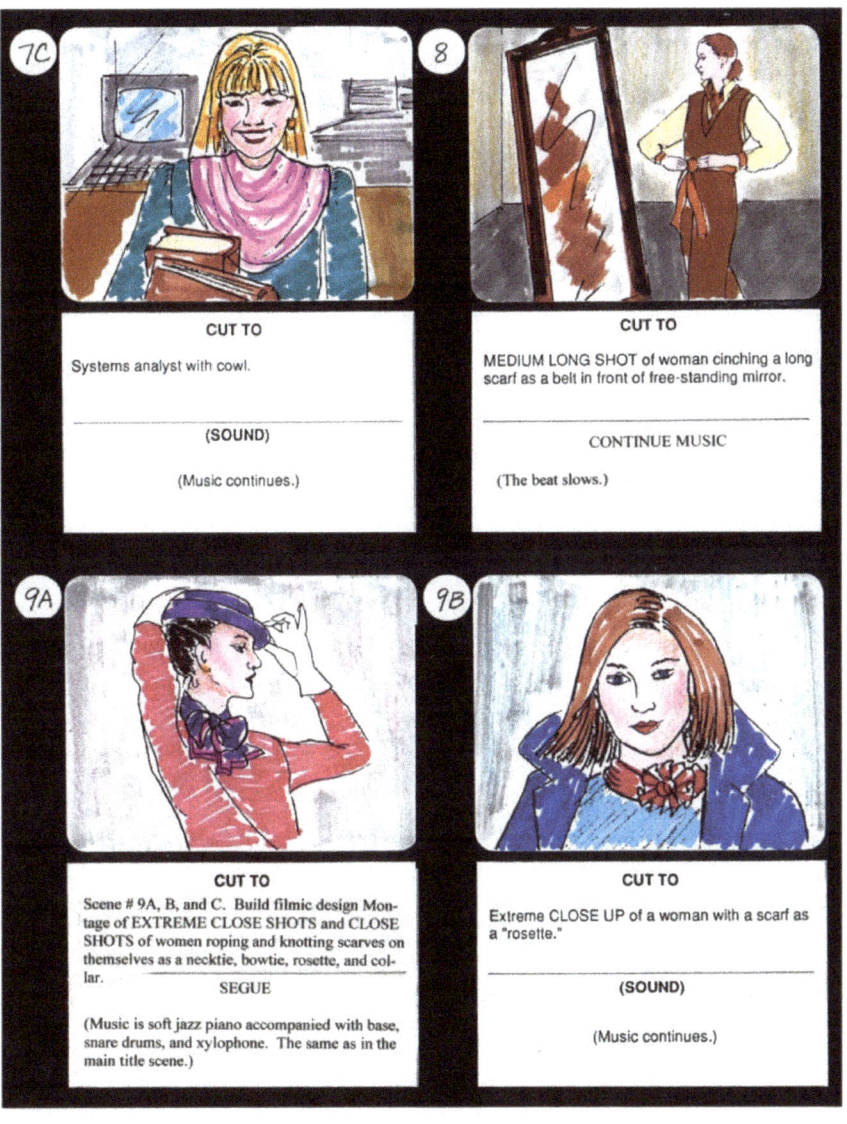

7C

CUT TO

Systems analyst with cowl.

_____

(SOUND)

(Music continues.)

8

CUT TO

MEDIUM LONG SHOT of woman cinching a long
scarf as a belt in front of free-standing mirror.

_____

CONTINUE MUSIC

(The beat slows.)

9A

CUT TO

Scene # 9A, B, and C. Build filmic design Montage of EXTREME CLOSE SHOTS and CLOSE
SHOTS of women roping and knotting scarves on
themselves as a necktie, bowtie, rosette, and collar.
_____
SEGUE

(Music is soft jazz piano accompanied with base,
snare drums, and xylophone. The same as in the
main title scene.)

9B

CUT TO

Extreme CLOSE UP of a woman with a scarf as
a "rosette."

_____

(SOUND)

(Music continues.)

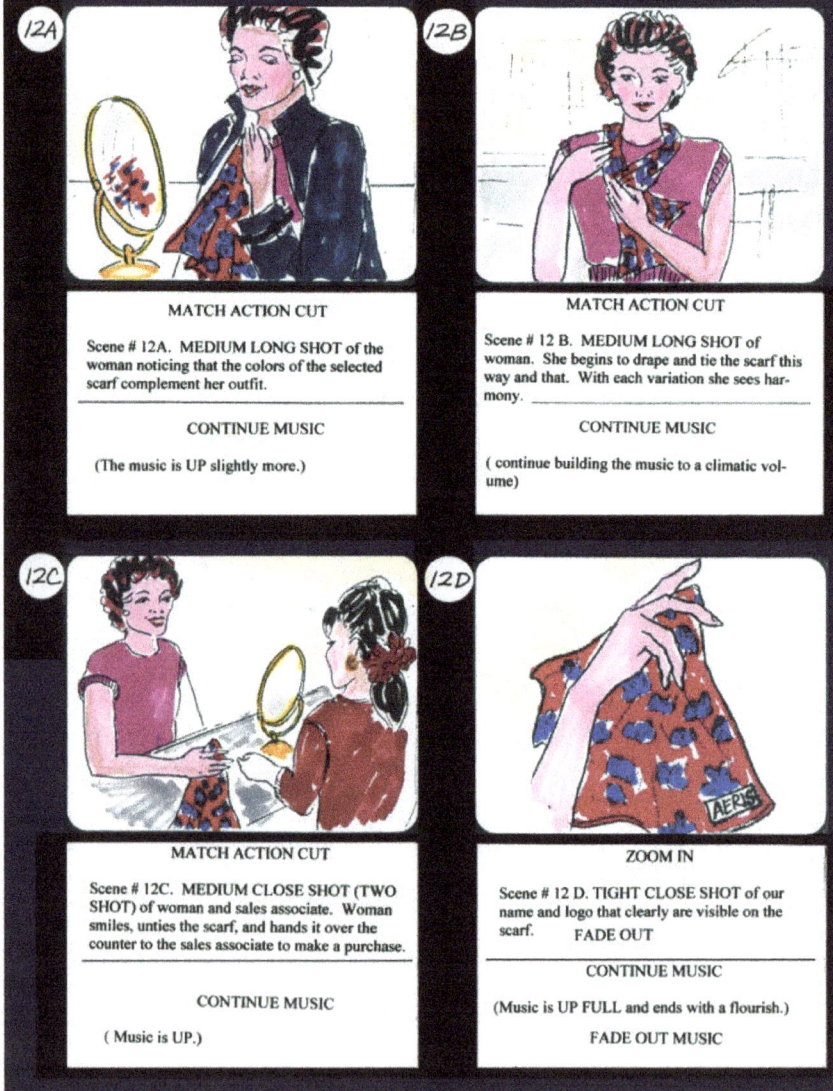

12A   MATCH ACTION CUT

Scene # 12A. MEDIUM LONG SHOT of the woman noticing that the colors of the selected scarf complement her outfit.

CONTINUE MUSIC

(The music is UP slightly more.)

12B   MATCH ACTION CUT

Scene # 12 B. MEDIUM LONG SHOT of woman. She begins to drape and tie the scarf this way and that. With each variation she sees harmony.

CONTINUE MUSIC

(continue building the music to a climatic volume)

12C   MATCH ACTION CUT

Scene # 12C. MEDIUM CLOSE SHOT (TWO SHOT) of woman and sales associate. Woman smiles, unties the scarf, and hands it over the counter to the sales associate to make a purchase.

CONTINUE MUSIC

(Music is UP.)

12D   ZOOM IN

Scene # 12 D. TIGHT CLOSE SHOT of our name and logo that clearly are visible on the scarf.   FADE OUT

CONTINUE MUSIC

(Music is UP FULL and ends with a flourish.)

FADE OUT MUSIC

Appendix 4

Information Teleplay

Gambling Addiction and the Family
(a script segment)

Jack Walker
Los Angeles, Calif.

EXT. SUBURBAN NEIGHBORHOOD - DAWN

Gray light and murky shadows. The ghostly, insubstantial hour between night and day. JANE trudges up the sidewalk, a flimsy windbreaker zipped over her pink waitress uniform.

> NARRATOR (VOICE-OVER)
> Addicts may spend days and weeks
> on a binge, gambling away their
> homes, their savings, even their families.

She walks up the front steps of a shabby tract house and quietly unlocks the front door.

INT. JANE'S HOUSE - DAWN

Cheap, mismatched furniture. Tiptoeing inside, Jane is surprised by a half-deflated toy balloon hovering behind the front door. She sees a sagging "Happy Anniversary" banner overhead.

> JANE
> (under her breath)

Damn.

Jane peers into the living room, sees her husband RICK snoring on the threadbare recliner.

Jane moves softly to Rick's side. She reaches for the jacket draped across his midsection, but instead of tucking him in, she rummages through the pockets. Her determined efforts awaken Rick.

RICK

You missed the party.

Startled, Jane makes a half-hearted attempt to hide the twenty-dollar bill she has found.

RICK

Go ahead, take it.

Ashamed, she sets the twenty on the coffee table.

JANE

Something came up. I had to cover for Cherise.

RICK

I called. You left work at eight.

JANE

I know, but I had things to do.

RICK

Alene was here. And Marty and Bernice. My folks, too. I kept telling them you were on the way but they stopped believing me eventually and let themselves out.

JANE

I was getting you a present.

Rick looks skeptical.

JANE

It's the truth. I mean, it was the truth.

RICK

You were gambling.

JANE

I didn't mean to. I borrowed a few bucks from Cherise. For your present, that's that's the only reason she was going to loan me the money. For that briefcase you'd been talking about. But, even with her money, I didn't have enough. So I thought maybe I could turn it into a little extra.

RICK

At the casino.

JANE

I doubled it right away, I was so hot. I should've come right home, but it would've been nuts to walk away when the cards were saying stay. I mean, I could've bought you the briefcase and fixed the fridge and gotten the TV out of hock.

Appendix 5

Split-Page with Voice-Over Narration

Desert Stewardship
(opening scene)

Produced for Naval Weapons Center
China Lake, California

Script by
Film Projects Branch

Desert Stewardship

CUT TO	SEGUE
Scene # 1. EXTERIOR, Mojave Desert at China Lake, California.	(NARRATOR)
LONG SHOT of Joshua Flat. Coso Peak is in the background. In the foreground is Black Spring. It is flowing. We see a forest of Josh-ua trees and other vegetation.	Most people think of the desert as a barren wasteland where very little can live or grow.
ZOOM IN	(Pause)
MEDIUM SHOT of Black Spring. We see the trickle of water coming from the spring. A Mojave ground squirrel sips the water.	Actually, such thoughts are far from the truth.
DISSOLVE	(MUSIC IN)
Scene # 2. EXTERIOR, various locations in the desert at China Lake. Build a filmic mon-tage of CLOSE SHOTS and MEDIUM SHOTS of various desert wildlife in action. We cross-cut with scenes having movement in opposite screen directions to emphasize the wide diversity of animal life. For example, sidewinder, partridge, mountain lion, coyote, weasel, deer, etc.	(Softly with a moderate western theme.) (NARRATOR) The desert is a delicately balanced ecosystem supporting over two hundred species of wild-life.

CUT TO	(NARRATOR)
Scene # 3. EXTERIOR. Continue the mon-tage. Focus on CLOSE SHOTS of the follow-ing animals	Several of the animals indigenous to this area are legally protected: . . .
(show and tell)	
bighorn sheep	. . . bighorn sheep, . . .
golden eagle	. . . golden eagle, . . .
desert tortoise	. . . desert tortoise, and . . .
Mojave ground squirrel	. . . Mojave ground squirrel.
CUT TO	(NARRATOR)
Scene # 4. EXTERIOR, Black Can-yon Area. Build a filmic montage of CLOSE SHOTS AND MEDIUM CLOSE SHOTS of the flora in the area.	(Pause) There are more than four hundred species of plants in this desert, ranging from . . .
CUT TO	(NARRATOR)
Scene # 5. EXTERIOR, Black Can-yon Area. Build filmic montage of CLOSE SHOTS of the various flora. CUT ON CUE to shots of pinyon pine and pickle weed	. . . pinyon pine and . . . . . . pickle weed . . .

Appendix 6

Teleplay and Split Page Combination
Pacific Frontier
(opening scene)

Produced for the Chief of Information, U.S. Navy

Erskin Gilbert
New York City

Opening narration by
Herman Melville

FADE IN   (MUSIC IN)

Scene 1. EXTERIOR. THE OCEAN.

The gently rolling sea as seen from the hangar deck of an aircraft carrier. As the ship pitches slightly, the scene becomes a CLOSE SHOT. As the sea rises, the scene becomes a LONG SHOT.

FADE IN and SUPERIMPOSE over Scene 1:

Title 1. THE UNITED STATES NAVY PRESENTS

FADE OUT Title 1.

Scene 1 continues to play.

SUPERIMPOSE Title 2, PACIFIC FRONTIER. Title 2 ZOOMS IN from infinity to full frame and holds.

FADE OUT Title 2.

Scene 1 continues to play. After a few seconds, the narrator begins.

                                        (NARRATOR)
                              There is one knows not what sweet mys-
                              tery about this sea whose gentle awful stir-
                              rings seem to speak of some hidden soul
                              beneath.

DISSOLVE TO

Scene 2. EXTERIOR. MAKAHA BEACH, HAWAII.

Huge, rolling breakers as they tumble close to the shore.

> (NARRATOR)
> It rolls the midmost waters of the world, the Indian Ocean and Atlantic being but its arms.

DISSOLVE TO

Scene 3. EXTERIOR, LA JOLLA BEACH.

CLOSE SHOT of a breaker as it smashes into the rocks on the beach.

> (NARRATOR)
> The same waves wash the moles of new-built California towns, but yesterday . . .

DISSOLVE TO

Scene 4. EXTERIOR, BIG SUR BEACH.

HIGH ANGLE SHOT of breakers as they roll onto the beach. Sea gulls are perched on rocks in the foreground.

> (NARRATOR)
> . . . planted by the recentest race of man . . .

DISSOLVE TO

Scene 5. EXTERIOR, REPULSE BAY, HONG KONG.

MEDIUM LONG SHOT of small Chinese junk with sails rigged putting out to sea. The day is overcast.

(NARRATOR)
. . . and lave the faded but still gorgeous
skirts of Asiatic lands, older than Abraham.

DISSOLVE TO

Scene 6. EXTERIOR, WAIALUA BAY, HAWAII.

LOW ANGLE SHOT from the water's surface, a huge breaker smashes over a coral reef. Spray flies.

(MUSIC UP)

Appendix 7

Shelton's Fundamental Verities of Information Motion-Media

I've listed below the fundamental elements of our motion-media profession. In essence, they are a summary of this book.

- We're in the communication profession rather than the film, video, or multimedia business.
- Actually, we're in the psychology business. Our task is to manipulate the minds of our audience members.
- The reason to produce a show (the problem to be solved) sometimes is not what it seems to be.
- Be absolutely sure you know precisely what the real problem is.
- Select the appropriate medium for the message and audience.
- Motion-media's primary communication advantage is that filmic design manipulates time and space.
- Film and video are linear; multimedia is interactive.
- Film and video are appropriate media for broad-based goals.
- Multimedia is appropriate for intensive training and learning.
- Only in multimedia can we get real-time feedback.
- Well-done audience analysis augurs well for successful communication.
- Filmic design is the grammar and syntax of motion-media—the key to encoding messages in motion-media communication.
- Encode motion-media messages in a style, content, and form with which the audience can identify.
- Engendering empathy in the audience is the key to communication.

- Motion-media are kinetic-visual media.
- Aural information must be kept to a minimum.
- Kinetic-visual and aural information in the right mix generate communication synergism—about 75 percent visual, 25 percent aural.
- Script design is kinetic-visual planning.
- Storyboard is the basic script format.
- Simpler and shorter shows are communication- and cost-effective.
- Five is about the maximum number of key points that should be in a linear show.
- Stilted scenarios usually fail.
- Large budgets do not equate to "better" shows.
- Information motion-media shows do not have to entertain to communicate.
- Creativity is basic and deliberately concealed.
- Entertainment and creativity all too often are noise that hinders communication.
- Talking heads are taboo, except in extraordinary cases.
- Clients aren't always right, but they pay the bills.
- You're responsible for your own professional development.
- Know the history and background of our profession—learn from the masters.
- Learn communication theory, psychology, and system analysis.
- Technical skills are developed on the job.
- Develop the critical eye.
- Become a catholic communicator.

Appendix 8

Quotable Quotes

Over the years, I've jotted down comments regarding our profession that I've heard associates say or that I've read in articles or in personal correspondence. Some are insightful, some are funny, and some strip away the veneer and expose the true essence of what's happening in our profession. I didn't fabricate any of these quotes. They are exactly as I heard or read them. On hearing a quotable quote, I'd write it down as soon as I could. I've listed some of these quotes below. They are in no particular order. Enjoy, reflect, and understand the intrinsic meaning in these quotes.

"First law of inverse correlations. The sophistication level of a project is inversely proportional to its importance." Gary Glendening, media producer, quoted in *Audio-Visual Communications,* November 1981.

"The standard notion is that for full appreciation, films and plays demand the willing suspension of disbelief on the part of the viewer." Charles Champlin, art critic, *Los Angeles Times,* 3 October 1980.

"This show is straight show and tell. Trouble is, you can't see anything." The late John Dunker, filmmaker and former associate, commenting on a show submitted for judging in an international film competition. In this film, the vast majority of the information was in the narration. Kinetic visuals were almost irrelevant.

"Anytime there was anything exciting happening [aerial combat], my combat photographers were either out of film or picking their noses." The late General Curtis E. LeMay, USAF, former commander of the 20th Air Force (B-29 bombers in the Pacific Theater) and later commander of the U.S. Air Force's Strategic Air Command. General LeMay made this comment to me in my office, 27 July 1977, after screening my film *299 Foxtrot*. His comment expressed his frustration with his film people who produced the mediocre documentary film, *The Last Bomb* (the story of the B-29 raids on Japan). LeMay wanted his film to compete with and overshadow William Wyler's famous documentary, *The Memphis Belle*, the story of the B-17's 25th mission—this time over the heavily defended port city of Wilhelmshaven.

"We're not dealing with the truth here [in this film]. What we're dealing with is an illusion [of the truth]." Anonymous filmmaker and former associate in a discussion with me on how a certain sequence in a film should be edited to overcome the shortcomings in footage needed to cover the information that had to be communicated.

Charles Kettering, physicist, summing up his lifetime experience with evaluation committees: "Their reaction is to seek the wrongness: to ignore 90 percent of the rightness." Quoted by Cap Palmer in his column "Open Mike," in *Film World and A-V News,* February 1965, p. 270.

"Let's have everybody's input so there won't be any complaints." sarcastic comment by Sam Stalos, editor and columnist, regarding review committees, in his column "Stalos: AV Shrink," *Audio-Visual Direction,* January 1984, p. 53.

"Anything can happen in the world of make-believe. The **silence** is the most important part of our show" (emphasis added). Hedda Sharapan, associate producer of Family Communications, Inc.'s *Mister Roger's Neighborhood,* on accepting the Distinguished Technical Communication Award at the Society for Technical Communication's annual conference, Pittsburgh, Pennsylvania, 21 May 1981.

"This film is a variety of scenes to complement the narration." The late Everett Baker, filmmaker, said this to me on screening his film in interlock to justify the show's lack of filmic design.

"The resulting proliferation of documents, addenda, declarations and self-supporting regulations vomit forth in [an] effort [that] is without merit, logic, or adequate word description." Jack Williamson, independent filmmaker, responding to the "Request for Bid" package of documents for a government film. "No Bid," *Business Screen,* May/June 1974.

"It [television] is adored and vilified. It's feared by some as a mind-altering drug or a subliminal messenger, ridiculed by others as a trivial boob tube, an idiot box; a vast, gray, barren, parched wilderness." Howard Rosenberg, television critic, in his column "50 years of TV: A Wasteland and a Wonder," *Los Angeles Times,* 10 April 1989, VI-1.

"Film is a bisensual medium." Tony Gurria, instructor at the London Film School, to me in a private conversation at the Society of Motion Picture and Television Engineers Conference, Los Angeles, California, December 1975.

"That (television), in short, is responsible for the downfall of civilization. We're in the midst of a new 'Dark Age.'" Ian Mitroff, professor of business policy, University of Southern California, in "False Images Lead Us Back to Dark Ages," *Los Angeles Times,* 11 October 1989, page 2, Section B7.

Does this metaphoric quote apply to your career? "The best scene is always at the bottom of the trim barrel." Chuck Bodwell, owner of Filmline Productions in Los Angeles, to me at an interlock screening of his film at the Naval Weapons Center, China Lake, California, November 1979.

"You simply can't cheat on a one-projector show and pull it off." Jean O'Neal, CEO of Corporate Image, Des Moines, Iowa, in personal correspondence to me, 4 March 1981. Jean was discussing multi-image shows that had lots of "creativity" but little communication. Such shows employed from two to thirty or forty 35mm slide projectors.

"Television has seduced us away from reality, deprived us of self-aware-ness and growth, given us ersatz excitement and vicarious adventure, fan-tasized our sex life, brutalized our consciousness with violence, stolen our time, homogenized us and turned us into zombies, couch potatoes, slobs, and mindless consumers." The late Jack Smith, columnist, *Los Angeles Times.* 20 February 1989. V-1.

"No navy training film never taught nobody nothing." Lieutenant Commander M. R. McClure, USN, production supervisor, Naval Photographic Center, Washington, D.C., April 1965. Said to me in disgust after we'd screened the first-answer print of a contractor-produced film for the navy. The film was totally inept and was reflective of the total scope of the "training" films then being produced.

An exceptionally talented filmmaker and former associate submitted his script to me for review. The film's subject dealt with a complicated research and development project. The meaning of several lines of narration was unclear. I asked my associate to explain this narration. In a burst of unintentional can-dor, the filmmaker replied with the ultimate faux pas, "I don't know what I'm talking about." Naval Weapons Center, China Lake, California, 10 May 1985.

In a stinging rebuke regarding television viewing, Derrick Z. Jackson wrote, "Television viewing has become a total public health menace. Brain rot from toxic, stereotyped shows." Derrick Z. Jackson, "NAACP Fight Is a Tired Rerun," *Boston Globe,* 29 August 2001.

Not all is lost. Even some of Hollywood's giants recognize the pitfalls of technology over story. For example, Robert Zemeckis, a technology-effects aficionado, notes that the true storyline comes first and foremost over technique. "I think it's the only way that a movie ultimately works under any circumstance. The technology is only there to **serve the story.** The story can never serve the technology" (emphasis added). Quoted in *Magazine,* John Zollinger's article "Giving Back to the Future," *USC Trojan Family*, Autumn 2001.

Lastly, Steven Spielberg said, when discussing the thread that ties all of his films together, "It (the script) has got to be a good story: compelling, one that will engage the audience." Quoted in "Profile: Steven Spielberg Talks about Technology and Storytelling." Frank Barron and Margie Barron, *UPDATE,* January 2004, p.3.

Finally, "The play's the thing," William Shakespeare, *Hamlet*, Act 2, Scene 2.

Appendix 9

Exercises

Now that you've completed this book and are ready to hone your motion-media script-designing skills, peruse the following real-world communication problems. Try to develop scripts that will solve (or at least alleviate) these communication problems. I've not included all the details of each problem. It's your challenge to develop those details that are missing as you complete the Communication Analysis and Motion-Media Preproduction Plan.

Completion of this plan is a crucial step in your solution of any problem. Think each question through carefully and respond with a cogent response. Next, outline your script in narrative form. Then try your hand at a storyboard—stick figures and rough drawings are okay. The goal here is to develop your filmic design for the solution to the problem. Finally, develop the script in one of the formats I've shown in the appendices. I suspect that you'll need several drafts before you're satisfied. Do your best.

Communication Problems

Problem 1: To Motivate. Your client is the manager of a huge warehouse complex. Warehouses are located in several western states. This manager hired an IT group to install a new computer system: it tracks purchases, inventory, sales, shipping, billing, etc. First-level and mid-level managers are reluctant to use this new computer system. They're comfortable with the ol' way that they know and that always works. Your task is to design a motion-media show that will motivate the managers to use the new computer system.

Problem 2: To Sell: The sales of your client's high-end computers are falling. A new computer company is cutting into your client's markets. Your client's computer and the competitor's are about equal in cost, capability, and reliability. Design a motion-media show that will boost sales to Fortune 500 companies, governments, and universities.

Problem 3: To Persuade. Student drug abuse, alcoholism, and vandalism are rampant in the local junior-high school. The principal has tasked you with designing a motion-media show that will assuage this problem. (Clearly, one show will not cure the problem.) What ancillary materials do you suggest be used with the show?

Problem 4: To Teach. The folks at the Senior Citizens' Center are trying to learn the basics of personal computing. They are having a difficult time. Design a motion-media show that will teach the senior citizens the basics of personal computing. What medium is best for this type of problem?

Problem 5: To Propagandize. The war is going badly for Zekistan. Their army suffers defeat after defeat. Casualties are high. Morale on the home front is low and headed lower. Design a motion-media show that will boost morale of the general public on the home front.

Problem 6: To Report. Physicists at Acme University have discovered the elusive and long-sought gravity wave—the last of nature's forces to be discovered. Such a discovery revolutionizes science and leads to the development of the "unified field theory." Design a motion-media show that will highlight this discovery to the scientific community. The peer-reviewed, mathematical paper will be published in two months in the *Acme Scientific Journal.*

Problem 7: To Persuade. Some animal-rights activists are opposed to using animals for any kind of medical research. Design a motion-media show that will cause some of these activists to change their minds or, at minimum, to reconsider their position.

Problem 8: To Inspire. Students in the English Poetry 101 class couldn't care less, and many are failing. This class is a required core class for all entering freshmen at Acme University. Design a motion-media show that will inspire the students to appreciate, become involved with, and thus learn or at least tolerate English poetry.

Problem 9: To Recruit. Minority representation in professional jobs at the Department of Defense poison gas and biological warfare research and development center is too low. A federal court order tasks the center with increasing minorities in the professions by 12.5 percent. This goal must be reached within two years. The center is located in a remote desert region of Nevada, 250 miles from the nearest town. It's hot in the summer and cold in the winter. Design a motion-media show that will recruit minority professionals for long-term employment at the center.

Problem 10: To Inform. Traffic in the center of town is a nightmare. To alleviate this situation, the city council is planning to eliminate all street parking and to make all streets one-way in this area. The council anticipates intense opposition from merchants, the public, and various special-interest groups. Members of the city council are to face the voters in the upcoming November elections. Design a motion-media show that will engender a positive attitude in the opposition.

References

Chapter 1. The Message and Motion-Media

1. Alvin Toffler, *Future Shock,* New York: Bantam Books, 1972, p. 166.
2. Everett L. Jones, "Subject A: The Class Increases," *Los Angeles Times*, 8 December 1984.
3. Mike Wallace, "Would CBS' Mike Wallace Switch Off His Own Show?" *Los Angeles Times,* 9 March 1976.
4. Charles R. Wright, *Mass Communications,* New York: Random House, 1959.
5. Reed H. Blake and Edwin O. Haroldsen, *A Taxonomy of Concepts in Communication,* New York: Hastings House, 1975.
6. Paul Lazerfield and Robert Merton, "Mass Communication, Popular Taste, and Organized Social Action," in *The Communication of Ideas: A Series of Addresses,* ed. Lyman Bryson, New York, Institute for Religious and Social Studies, 1948.
7. Rudolf Arnheim, *Visual Thinking,* Berkeley: University of California Press, 1969.
8. Marshall McLuhan, *Understanding Media: The Extensions of Man,* 2nd ed., New York, New American Library, 1964.
9. Arnheim, *Visual Thinking.*
10. Charles F. Hoban and Edward B. Van Ormer, *Instructional Film Research (Rapid Mass Learning) 1918 to 1950,* University Park, Penn., Pennsylvania State College, 1951.

Chapter 2. Motion-Media in the Communication Society

1. Nicholas Negroponte, "The Future of TV and the Electronic Communication Media," speech delivered to the International Television Association's 23rd International Conference, Boston, 30 May 1991.

2. Charles S. Steinberg, *Mass Media and Communication*, 2nd ed., New York: Hastings House, 1972.

3. John Fiske, *Introduction to Communication Studies*, 2nd ed., New York: Routledge, 1990.

4. Father Louis Reile, Society of Mary, "Movies: Major Prophet of Our Time," *Gold and Blue*, San Antonio: St. Mary's University, October 1978, p. 2.

5. Arnheim, *Visual Thinking.*

Chapter 3. Some Thoughts on Information, Communication, and Meaning

1. David K. Berlo, *The Process of Communication*, New York: Holt, Rinehart and Winston, 1960, p. 30.

2. Fiske, *Introduction to Communication Studies*.

3. C. L. Shannon and Warren Weaver, "The Mathematical Theory of Communications," *Bell System Technical Journal*, July/October 1948.

4. Jeremy Campbell, *Grammatical Man: Information, Entropy, Language, and Life*, New York: Simon and Schuster, 1982.

5. James Gleick, *Chaos: Making a New Science*, New York: Penguin Books, 1987, pp. 257-58.

6. If your interest is piqued by entropy and you'd like more information, I recommend reading Campbell, *Grammatical Man*; Gleick, *Chaos*; and Hans Christian von Baeyer, *Maxwell's Demon: Why Warmth Disperses and Time Passes*, New York: Random House, 1998.

Chapter 4. Creativity May Not Equal Communication

1. Lou Clement, *Acquisition of Motion Pictures and Videotape Productions*, Washington, D.C.: Office of the Assistant Secretary of Defense for Public Affairs, 1984, p. 3.

2. Ed Gray, "Better Business Through Creative Freedom," *Audiovisual Communications*, June 1981, p. 50.

3. David MacLeod, "Creativity: The Essence of Communication," *COMMTEX International NAVA/AECT Daily,* 22 January 1984, p. 1.

4. Charles (Cap) Palmer, "A Personal Philosophy About 'Business Films,'" *Exit Lines: A Personal Philosophy About Factual Films,* Ames: Iowa State University, 1982, p. 3.

5. Barbara Saltzman, "Sinatra Credits German 'Stranger,'" *Los Angeles Times,* 11 September 1982, IV-3.

6. Jean O'Neil, personal correspondence with the author, 4 March 1981.

7. John Grierson, "The Documentary Idea, 1942," in *Grierson on Documentary,* ed. Forsyth Hardy, New York: Praeger, 1975, p. 257.

8. Eleanor Wright, "How Creativity Turns Facts into Usable Information," *Technical Communication,* 1985, p. 29.

9. Charles F. Hoban Jr., "The State of the Art of Film in Instruction: A Second Look," *Audiovisual Instruction,* April 1975, pp. 30-34.

10. Robert Davis, "The Art of Creation," *Audio-Visual Communications,* November 1981, p. 26.

Chapter 5. Information Motion-Media

1. John Grierson, "The E.M.B. Film Unit," in *Grierson on Documentary*, p. 165.

2. John Grierson, "A Movement Is Founded," ibid., p. 185.

3. J. Walter Klein, *The Sponsored Film,* New York: Hastings House, 1976.

4. Joseph V. Mascelli, *The Five C's of Cinematography: Motion Picture Filming Techniques,* 1965, Los Angeles: Silman James Publishers, 1998.

5. Hoban and Van Ormer, *Instructional Film Research,* pp. 8–30.

6. Lloyd Engler, "A Commercial Message," Audio-Visual Communications, April 1980: p. 45.

7. Charles (Cap) Palmer, "Single Concept Comes of Age," *Educational Screen and Audiovisual Guide,* December 1963: p. 17.

8. Lloyd Harvey, "Thoughts on Making Short Films and Other Pleasures," *Business Screen,* September/October 1975: p. 27.

Chapter 6. Linear Media: Video and Film

1. Don Dillon, "Hearing, Seeing, Reading," survey by the National Audiovisual Association, reported in the thirtieth anniversary issue of *Crowley Commentary Newsletter,* Toronto: Crawley Film Ltd., 1980, quoted in *Industrial Communication Council Newsletter,* July 1980.

2. Wilson B. Key, *Subliminal Seduction,* New York: Plume and Meridian Books, 1973, p. 163.

3. Hoban and Van Ormer, *Instructional Film Research.*

4. Karel Reisz, *The Techniques of Film Editing,* New York: Visual Arts Books, 1953, p. 265.

5. John Grierson in *Grierson on Documentary.*

6. Sergei Eisenstein, with Jay Leyda, *The Film Sense,* New York: Harcourt Brace, 1942, p. 4.

Chapter 7. The False Reality of Motion-Media

1. Brian Henderson, *A Critique of Film Theory,* New York: Dutton, 1980, pp. 19-21.

2. Siegfried Kracauer, *Theory of Film: The Redemption of Physical Reality,* London: Oxford University Press, 1978; André Bazin, "The Ontology of the Photographic Image," in *What Is Cinema? Essays,* ed. and trans. Hugh Grey, Berkeley: University of Calif. Press 1968, p. 9.

3. Jay Leyda, *Kino,* New York: Collier Books, 1973, p. 162.

4. Paul Rotha, *The Film Till Now,* London: Spring Books, 1967, p. 122.

5. Paul Rotha, *Documentary Diary,* New York: Hilland Way, 1972, p. xvi; Grierson, *Grierson on Documentary,* p. 13.

6. Allardyce Nicoll, "Film Reality: The Cinema and the Theatre," in *Film: An Anthology,* ed. Daniel Talbot, Berkeley: University of California Press, 1972, p. 49.

7. Leon Barsacq, *Caligari's Cabinet and Other Grand Illusions,* New York: New American Library, 1976, p. 46.

8. Grierson, *Grierson on Documentary,* p. 13.

9. Rotha, *The Film Till Now,* pp. 113–15; Klein, *The Sponsored Film,* p. 12.

10. S. Martin Shelton, "What Happens When the Lights Come On?" *Proceedings of the 27th International Technical Communication Conference, May 14–17, 1980, Minneapolis, Minnesota,* Washington, D.C.: Society for Technical Communication, 1980, p. G-81.

Chapter 8. Documentary Film: A Learning Tool

1. John Grierson, quoted in *Grierson On Documentary*, p. 16.

2. Rotha, *Documentary Diary,* p. xv; John Grierson, quoted in Rachel Low, *Documentary and Educational Films of the 1930s,* London: George Allen & Unwin, 1979, p. 176; Roger Manvell, *The Film and the Public,* Harmondsworth, England: Penguin Books, 1955.

3. John Grierson, "The Course of Realism," in *Grierson on Documentary,* p. 207.

4. John Grierson, "First Principles of Documentary," pp. 146–47.

5. Ibid.

6. Rotha, *The Film Till Now,* p. 123.

7. Ivor Montagu, *Film World: A Guide to Cinema,* Baltimore: Penguin Books, 1964, p. 285.

8. Robert L. Snyder, *Pare Lorentz and the Documentary Film,* Norman: University of Oklahoma Press, 1968, p. 3.

9. John Grierson, "Documentary: A World Perspective," in *Grierson on Documentary,* p. 385.

Chapter 9. Introduction to Multimedia

1. Dawn Stover, "Hypermedia," *Popular Science* 234, No. 5 (May 1989): pp. 122–24, 160.

2. Gillen Interactive Group, Inc., "Creating Your First Compact Disc Interactive Project," press release, Laytonsville, Maryland, November 1990; Nathan Kolowski, "Development of Effective Interactive Instruction Materials," *Instruction Delivery Systems 1,* No. 2, March/ April 1987: p. 19.

3. Gregory L. Adams, "Why Interactive," Multimedia and Videodisc Monitor 10, No. 3 (1992): pp. 20–24.

4. Clara Lazzareschi, "AT&T Enters Race to Offer Interactive TV," *Los Angeles Times,* 6 May 1993, D1–D3.
5. Clara Lazzareschi, "TV Till You," *Los Angeles Times Magazine,* 16 May 1993, pp. 12–16.

Chapter 10. Multimedia Flowchart
None.

Chapter 11. Stating Your Objectives, or, What's This Show About?
None.

Chapter 12. *Who's Our Audience?*
1. Melvin L. DeFleur, *Sociology: Man in Society,* Glenview, Ill.: Scott, Foresman, 1974, p. 419.
2. Melvin L. DeFleur, *Theories of Mass Communications,* New York: 1961, p. 121.
3. Joseph T. Klapper, *The Effects of Mass Communication,* New York: Free Press, 1960.
4. Bernard Berelson and Gary A. Steiner, *Human Behavior,* New York: Harcourt, Brace and World, 1964.
5. Klapper, *The Effects of Mass Communication.*
6. Hoban and Van Ormer, *Instructional Film Research,* pp. 8-1 to 8-43.
7. McLuhan, Marshall, *Understanding Media,* p. 24.
8. Hoban and Van Ormer, *Instructional Film Research,* pp. 3–11.

Chapter 13. Creating a Communication Analysis and Motion-Media Communication Plan
None.

Chapter 14. Filmic Design
None.

Chapter 15. An Information Motion-Media Writer Should Be a Script Designer
None.

Chapter 16. Scripting for Information Motion-Media
1. Donna Matrazza, *The Corporate Scriptwriting Book,* Boston: Medial Concepts Press, 1976.

Chapter 17. Marty's Contrary Principles of Script Design
None.

Chapter 18. Guidelines for Writing Narration and Dialogue
1. Marilyn Morgan, personal correspondence with the author, 18 October 2000.

Chapter 19. The Talking Head
None.

Chapter 20. The Sound Track: An Integral Part of Information Motion-Media
1. Bela Balazs, *Theory of Film,* New York: Dover, 1970, pp. 205–06.

Chapter 21. Distribution
None.

Chapter 22. Our Client Isn't the Enemy
1. Producers' Guild Ethics Committee, "The Work Ethic," *Audiovisual Communications* (June 1987): p. 26.
2. John Grierson, "The Malaise of Disillusionment," in *Grierson on Documentary,* p. 358.
3. Charles (Cap) Palmer, "Open Mike," *Film World and A-V News* (February 1965): p. 10.
4. Susan Sherman and Jay Libby, "Meeting the Client's Needs," *Video Systems* (February 1986): p. 44.
5. Raul Da Silva and Richard H. Rogers, *The Business of Filmmaking,* Rochester: Eastman Kodak Company, 1978, p. 6.
6. Alan Amenda, "A Gallery of Audio-Visual Hang-Ups," *Audiovisual Communications* (December 1975): p. 25.

7. Thomas R. Carlisle, "Point of View," *Industrial Cine Video* (October 1984): p. A34.

8. Murry C. Christensen, "What Every Client Should Know," *Audiovisual Communications* (April 1982): p. 38.

9. Mark Pahuta, "Perceptions of the In-House Film Unit by an Audiovisual Apprentice," *Proceedings, 25th International Technical Communication Conference May 10–13, 1978*, (Washington, D.C.: Society for Technical Communication, 1978), p. 123.

10. Sam Stalos, "Stalos: AV Shrink," *Audiovisual Directions,* (January 1984): p. 53.

11. Ralph Metzner, "Video Vanities," *Audiovisual Communications,* (November 1986): p. 15.

12. Palmer, *Exit Lines,* p. 3.

13. John Grierson, "Progress and Prospect," in *Grierson on Documentary,* p. 356.

Bibliography

Note: I recommend those titles marked with an asterisk for reading soonest.

Adams, Gregory L. "Why Interactive." *Multimedia and Videodisc Monitor* 10, No. 3 (1992): pp. 20 to 24.

Alber, A. F., *Interactive Computer Systems: Videotext Multimedia*. New York: Plenum, 1989.

Amenda, Alan. "Audiovisual Writing." In *Jobs for Writers*. New York: Writer's Digest Books, 1984. "A Gallery of Audio-Visual Hang-Ups." *Audiovisual Communications* (December 1975).

Arijon, Daniel. *Grammar of the Film Language*. London: Focal Press, 1976.

Arnheim, Rudolf. *Visual Thinking*. Berkeley: University of California Press, 1969.

*Atkins, P. W. *The Second Law*. New York: Scientific American Library, 1984.

Audio Visual Market Place. New York: R. R. Bowker, published annually.

Balazs, Bela. *Theory of Film*. New York: Dover, 1970.

Barnouw, Erik. *Documentary: A History of the Non-Fiction Film*. New York: Dell, 1974.

Barsacq, Leon. *Caligari's Cabinet and Other Grand Illusions*. New York: New American Library, 1976.

Barsam, Richard Meran, ed. *Nonfiction Film Theory and Criticism*. New York: Dutton, 1976.

Bazin, André. "The Ontology of the Photographic Image." In *What Is Cinema?: Essays,* edited and translated by Hugh Grey. Berkeley: University of California Press, 1968.

Beraman, Robert E., and Thomas V. Moore. *Managing Interactive Video and Multimedia Projects*. Boston: Knowledge Industry Publications, 1990.

Berelson, Bernard, and Gary A. Steiner. *Human Behavior*. New York: Harcourt, Brace and World, 1964.

Berger, Jeff. *The Desktop Multimedia Bible*. New York: Addison Wesley, 1993.

Berlo, David K. *The Process of Communication*. New York: Holt, Rinehart and Winston, 1960.

Blake, Reed H., and Edwin O. Haroldsen. *A Taxonomy of Concepts in Communication*. New York: Hastings House, 1975.

Blakefield, Bill. *Documentary Film Classics*. Washington, D.C.: National Audiovisual Center, 1985.

Blum, Richard A. *Television Writing*. New York: Focal Press, 1984.

*Campbell, Jeremy. *Grammatical Man: Information, Entropy, Language, and Life*. New York: Simon and Schuster, 1982.

Carlisle, Thomas R. "Point of View," *Industrial Cine Video*, (October 1984).

Christensen, Murry C. "What Every Client Should Know." *Audiovisual Communications* (April 1982).

Clement, Lou, *Acquisition of Motion Pictures and Videotape Productions*. Washington, D.C.: Office of the Assistant Secretary of Defense for Public Affairs, 1984.

Corliss, Richard. "Robert Flaherty: The Man in the Iron Myth." In *Nonfiction Film Theory and Criticism*, edited by Richard Meran Barsam. New York: Dutton, 1975.

Cotton, Bob, and Richard Oliver. *Understanding Hypermedia: From Multimedia to Virtual Reality*. London: Phaidon, 1990.

Da Silva, Raul, and Richard H. Rogers. *The Business of Filmmaking*. Rochester: Eastman Kodak Company, 1978.

Davis, Robert. "The Art of Creation." *Audio-Visual Communications* (November 1981).

DeFleur, Melvin L. *Sociology: Man in Society*. Glenview, Ill.: Scott, Foresman, 1974
. ---. *Theories of Mass Communications*. New York: David McKay, 1961.

Desmarais, Norman. *Multimedia on the PC: A Guide for Information Professionals*. New York: Meckler, 1993.

Dillon, Don. "*Hearing, Seeing, Reading*." Survey by the National Audiovisual Association reported in the Thirtieth Anniversary issue of "Crowley

Commentary Newsletter" (Toronto: Crawley Film Ltd., 1980), quoted in "Industrial Communication Council Newsletter" (July 1980).

*Dillon, Patrick M., and David C. Leonard. *Multimedia and the Web from A to Z.* Phoenix: Oryx Press, 1999.

Eastman Kodak Company. *How To Be a Knockout With AV!* Kodak Publication 8-84. Rochester, N.Y.: Eastman Kodak, 1984.

---. *Movies with a Purpose.* Kodak Publication No. V1-13. Rochester, N.Y.: Eastman Kodak, 1976.

Eisenstein, Sergei, with Jay Leyda. *Film Form.* New York: Harcourt Brace, 1949.

---. *The Film Sense.* New York: Harcourt Brace, 1942.

Emonds, Robert. *Scriptwriting for the Audio-Visual Media.* 2nd ed. New York: Columbia University, Teachers College Press, 1984.

Engler, Lloyd. "A Commercial Message." *Audio-Visual Communications* (April 1980).

Feldman, Tony. *An Introduction to Digital Media.* New York: Routledge, 1997.

Fiske, John. *Introduction to Communication Studies.* 2nd ed. New York: Routledge, 1990.

Gayeski, Diane M. *Interactive Toolkit.* Stoneham, Mass.: Focal Press, 1989.

Gayeski, Diane, and David Williams. *Interactive Media.* Englewood Cliffs, N.J.: Prentice-Hall, 1985.

Gillen Interactive Group, Inc. "Creating Your First Compact Disc Interactive Project." Press release, Laytonsville, Maryland, November 1990.

Gleick, James. *Chaos: Making a New Science.* New York: Penguin Books, 1987.

Gray, Ed. "Better Business Through Creative Freedom." *Audiovisual Communications* (June 1981).

Gregory, Mollie. Making Films Your Business. New York: Schocken, 1979.

*Grierson, John. *Grierson on Documentary,* edited and complied by Forsyth Hardy. New York: Praeger, 1975.

Grule, Katherine, ed. *Encyclopedia of Associations.* Detroit: Gale Research Company, current edition.

Harvey, Lloyd. "Thoughts on Making Short Films and Other Pleasures." *Business Screen* (September/October 1975).

Hayward, Stan. *Scriptwriting for Animation.* New York: Focal Press, 1977.

Henderson, Brian. *A Critique of Film Theory.* New York: Dutton, 1980.

Herdeg, Walter, ed. *Film & TV Graphics: An International Survey of the Art of Film Animation.* 2nd ed. New York: Hastings House, 1976.

Hilliard, Robert L. *Writing for Television and Radio.* 4th ed. Belmont, Calif.: Wadsworth, 1984.

Hoban, Charles F., Jr. "The State of the Art of Film in Instruction: A Second Look." *Audiovisual Instruction* (April 1975): 30–34.

*Hoban, Charles F., and Edward B. Van Ormer. *Instructional Film Research (Rapid Mass Learning) 1918-1950.* University Park, Penn.: Pennsylvania State College, 1951.

Imke, Steven. *Interactive Video Management & Production.* New York: Educational Technology Publications, 1991.

International Quorum of Motion Picture Producers. *Standard Motion Picture Production Contract.* Oakton, Virginia. IQ, 1984.

Iuppa, Nicholas V. *A Practical Guide to Interactive Video Design.* Stoneham, Mass.: Focal Press, 1984.

Jones, Everett L. "Subject A: The Class Increases." *Los Angeles Times,* 8 December 1984.

Key, Wilson B. *Subliminal Seduction.* New York: Plume and Meridian Books, 1973.

Klapper, Joseph T. *The Effects of Mass Communication.* New York: Free Press, 1960.

Klein, J. Walter. *The Sponsored Film.* New York: Hastings House, 1976.

Kolowski, Nathan. "Development of Effective Interactive Instruction Materials." *Instruction Delivery Systems* 1, No. 2 (March/April 1987).

Kracauer, Siegfried. *Theory of Film: The Redemption of Physical Reality.* London: Oxford University Press, 1978.

Lazerfield, Paul, and Robert Merton. "Mass Communication, Popular Taste, and Organized Social Action." In *The Communication of Ideas: A Series of Addresses,* edited by Lyman Bryson. New York: Institute for Religious and Social Studies, 1948.

Lazzareschi, Clara. "AT&T Enters Race to Offer Interactive TV." *Los Angeles Times,* 6 May 1993, pp. D1, D3. "TV Till You." *Los Angeles Times Magazine,* 16 May 1993, pp. 12–16.

Lee, Robert, and Robert Misiorowski. *Script Models: A Handbook for the Media Writer*. New York: Hastings House, 1984.

Leyda, Jay. *Kino*. New York: Collier Books, 1973.

*Leyda, Jay, and Zina Joynow. *Eisenstein at Work*. New York: Pantheon Books, 1982.

Lipton, Russell. *Multimedia Tool Kit*. New York: Random House, 1992.

Loomis, Andrew. *Creative Illustration*. New York: Viking Press, 1947.

*Lorentz, Pare. *FDR's Moviemaker: Memoirs and Scripts*. Reno: University of Nevada Press, 1992.

*---. *Lorentz on Film: Movies 1927 to 1941*. New York: Hopkinson and Blake Publishers, 1975.

*Low, Rachel. *Documentary and Educational Films of the 1930s*. London: George Allen & Unwin, 1979.

MacCann, Richard Dyer. *The People's Films*. New York: Hastings House, 1973.

MacLeod, David. "Creativity: The Essence of Communication." COMMTEX International NAVA/AECT Daily, 22 January 1984.

Maher, Charles. "3 Key Elements in All Contracts." *Los Angeles Times*, 10 September 1980.

Manvell, Roger. *Films and the Second World War*. New York: Dell, 1974.

---. *The Film and the Public*. Harmondsworth, England. Penguin Books, 1955.

*Mascelli, Joseph V., *The Five C's of Cinematography: Motion Picture Filming Techniques*. 1965. Los Angeles: Silman James Publishers, 1998.

Matrazza, Donna. *The Corporate Scriptwriting Book*. Boston: Media Concepts Press, 1976.

McLuhan, Marshall. *Understanding Media: The Extensions of Man*. 2nd ed. New York: New American Library, 1964.

Metzner, Ralph. "Video Vanities." *Audiovisual Communications* (November 1986).

Mezey, Phiz. *Multi-Image Design and Production*. Stoneham, Mass.: Focal Press, 1988.

Montagu, Ivor. *Film World: A Guide to Cinema*. Baltimore: Penguin Books, 1964.

Morley, John. *Scriptwriting for High-Impact Videos*. Belmont, Calif.: Wadsworth, 1992.

Naisbitt, John. *Megatrends*. New York: Warner Books, 1984.

Negroponte, Nicholas. "*The Future of TV and the Electronic Communication Media.*" Speech delivered to the International Television Association's 23rd International Conference, Boston, 30 May 1991.

Nicoll, Allardyce. "Film Reality: The Cinema and the Theatre." In *Film: An Anthology*, edited by Daniel Talbot. Berkeley: University of California Press, 1972.

Orbanz, Eva. *Journey to a Legend and Back*. Berlin: Edition Bolker Spiess, 1977.

Pahuta, Mark. "Perceptions of the In-House Film Unit by an Audiovisual Apprentice." *Proceedings, 25th International Technical Communication Conference May 10-13, 1978*. Washington, D.C.: Society for Technical Communication, 1978.

Palmer, Charles (Cap). "A Personal Philosophy About 'Business Films.'" In *Exit Lines: A Personal Philosophy About Factual Films*. Ames: Iowa State University, 1982.

---. "Open Mike." *Film World and A-V News* (February 1965).

---. "Single Concept Comes of Age." *Educational Screen and Audiovisual Guide* (December 1963).

Peter, Lawrence J. *The Peter Principle*. Toronto: Raymond Hull, Bantam Books, 1970.

Producers' Guild Ethics Committee. "The Work Ethic." *Audiovisual Communications* (June 1987).

Rabiger, Michael. *Directing the Documentary*. Boston: Focal Press, 1987.

Reid, J. Christopher, and Donald W. MacLennan, eds. *Research in Instructional Television and Film*. Catalogue No. 234-34041. Washington, D.C.: U.S. Government Printing Office, 1967.

Reile, Louis, SM. "Movies: Major Prophet of Our Time." *Gold and Blue* (San Antonio: St. Mary's University) (October 1978).

Reiser, Robert A., and Robert M. Gange. *Selecting Media for Instruction*. Englewood Cliffs, N.J.: Educational Technology Publications, 1983.

Reisz, Karel. *The Techniques of Film Editing*. New York: Visual Arts Books, 1953.

*Rifkin, Jeremy. *Entropy*. New York: Bantam Books, 1981.

Rosenberg, Victoria. *Guide to Multimedia*. New York: New Riders Publishing, 1993.

Rosenthal, Alan. *The New Documentary in Action: A Casebook in Film Making*. Berkeley: University of California Press, 1971.

Rotha, Paul. *Documentary Diary*. New York: Hilland Way, 1972.

---. *The Film Till Now*. London: Spring Books, 1967.

Rotha, Paul, and Richard Griffith. *Documentary Film*. 3rd ed. New York: Hastings House, 1952.

Saltzman, Barbara. "Sinatra Credits German 'Stranger.'" *Los Angeles Times*, 11 September 1982.

Shannon, C. L., and Warren Weaver, "The Mathematical Theory of Communications." *Bell System Technical Journal* (July/October 1948).

Shelton, S. Martin. "What Happens When the Lights Come On?" *Proceedings of the 27th International Technical Communication Conference, May 14-17, 1980*, Minneapolis, Minnesota. Washington, D.C.: Society for Technical Communication, 1980.

Sherman, Susan, and Jay Libby. "Meeting the Client's Needs." *Video Systems* (February 1986).

Snyder, Robert L. *Pare Lorentz and the Documentary Film*. Norman: University of Oklahoma Press, 1968.

Stalos, Sam. "Stalos: AV Shrink." *Audiovisual Directions* (January 1984).

Steinberg, Charles S., *Mass Media and Communication*. 2nd ed. New York: Hastings House, 1972.

Stephenson, Ralph, and J. R. Debrix. *The Cinema as Art*. Baltimore: Penguin Books, 1969.

Stover, Dawn. "Hypermedia." *Popular Science* 234, No. 5 (May 1989): pp. 122–124, 160.

*Sussex, Elizabeth. *Rise and Fall of the British Documentary*. Berkeley: University of California Press, 1975.

Swain, Dwight V. *Film Scriptwriting: A Practical Manual*. Boston: Focal Press, 1976.

---. *Scripting for the New AV Technologies*. 2nd ed. Stoneham, Mass.: Focal Press, 1991.

Toffler, Alvin. *Future Shock*. New York: Bantam Books, 1972.

*Tufte, Edward R. *Envisioning Information*. Cheshire, Conn.: Graphic Press, 1990.

---. *The Visual Display of Quantitative Information*. Cheshire, Conn.: Graphic Press, 1983.

---. *Visual Explanations*. Cheshire, Conn.: Graphic Press, 1997.

Ulrich's International Periodical Directory. New York: R. R. Bowker, current edition.

Vaughn, Tav. *Multimedia: Making It Work*. New York: Osborne/McGraw-Hill, 1993.

Von Baeyer, Hans Christian. *Maxwell's Demon: Why Warmth Disperses and Time Passes*. New York: Random House, 1998.

Wallace, Mike. "Would CBS' Mike Wallace Switch Off His Own Show?" *Los Angeles Times*, 9 March 1976.

Williams, Frederick. *The Communication Revolution*. Rev. ed. New York: New American Library, 1983.

Wolf, Rilla. *The Writer and the Screen*. New York: Morrow, 1974.

Wright, Charles R. *Mass Communications*. New York: Random House, 1959.

Wright, Eleanor. "How Creativity Turns Facts Into Usable Information." *Technical Communication (1985)*.

Zettl, Herbert. *Sight, Sound, Motion: Applied Media Aesthetics*. Belmont, Calif.: Wadsworth, 1973.

Index

About the Author

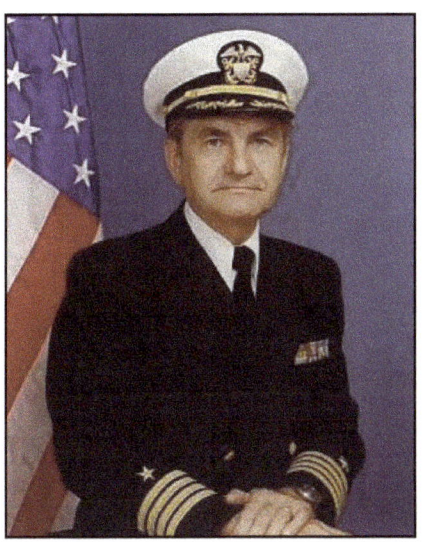

Captain S. Martin Shelton retired from active and reserve naval service with the rank of captain. He served in the Korean and Vietnamese Wars, and elsewhere as a combat motion-picture cameraman, photographic officer, and intelligence specialist. He has an extensive background in Far Eastern studies.

Shelton earned his Master of Arts Degree (Cinema) at the University of Southern California. He concentrated his studies on communications analysis, information theory, and the grammar and syntax of film scripting and production.

He has produced a host of information and documentary motion-media shows, winning over forty awards in national and international film competitions and festivals. His peers elected him a Fellow of both the Information Film Producers of America and the Society for Technical Communication. He served as the President of the Information Film Producers of America.

He has published extensively in trade magazines, peer-reviewed journals, and commercial publications. His professional book, *Communicating Ideas with Film, Video, and Multimedia,* garnered the Best of Show award in the Society for Technical Communication's Spotlight Publication Competition.

Shelton has posted more information about his work at sheltoncomm.com.

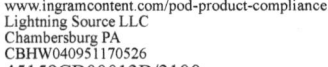